LIVING
WITH THE
EARTH

LIVING
WITH THE
EARTH

a manual for market gardeners

Volume 3

CREATING A MICROFARM

Perrine & Charles Hervé-Gruyer

PERMANENT PUBLICATIONS

Published in English by
Permanent Publications
Hyden House Ltd
13 Clovelly Road
Portsmouth
PO4 8DL
United Kingdom
Tel: 01730 776 582
international: +44 (0)1730
Email: enquiries@permaculture.co.uk
Web: www.permanentpublications.co.uk

Distributed in North America by Chelsea Green Publishing Company, PO Box 428, White River Junction, VT 05001 www.chelseagreen.com

Distributed in Australia by Peribo Pty Limited, 58 Beaumont Road, Mt Kuring-Gai, NSW 2080 Australia https://peribo.com.au

First published in French as *Vivre Avec La Terre*
© Actes Sud / Ferme du Bec Hellouin, 2019
Published in English in 2024 by Permanent Publications

English translation by John F Reynolds
English translation edited by Joni Praded

Designed by Two Plus George Limited, info@twoplusgeorge.co.uk

Printed in the UK by
SRP Ltd, Bittern Road, Sowton Industrial Estate, Exeter EX2 7LW

This product is made of material from well-managed FSC®-certified forests and from recycled materials and other controlled sources.

The Forest Stewardship Council ® (FSC) is a non-profit international organisation established to promote the responsible management of the world's forests. Products carrying the FSC label are independently certified to assure consumers that they come from forests that are managed to meet the social, economic and ecological needs of present and future generations.

British Library Cataloguing-in-Publication Data
A catalogue record for this book is available from the British Library

ISBN 978 1 85623 282 1

CHOU
CELERI
CHOU
DE
CHINE
HARICOT
NAVET
PANAIS
TOMATE
JEUNE
POUSSE

VOLUME I

PERMACULTURE, ECOCULTURE: INSPIRED BY NATURE

VOLUME II

FOOD CROPS AND FOREST GARDENS

VOLUME III

CREATING A MICROFARM

XI

PROPAGATING PLANTS

◄ *Collecting seeds is one of our priorities, but requires time and know-how. Although the work is intense, we always find time to harvest the easy-to-access seeds, like those shown here: sunflower, maize, coriander and peas.*

1

PLANT REPRODUCTION

After almost a century of industrial agriculture expansion, the Food and Agriculture Organization of the United Nations (FAO) now estimates that more than 75% of the world's crop biodiversity has been lost. Although many of us are unaware of it, this reality is quite visible on our supermarket shelves. Wherever the market economy has replaced subsistence farming, the agricultural offering is sadly familiar.

Blanche Magarinos-Rey[1]

For as long as agriculture has existed, cultivated plants have been multiplied by gardeners and farmers.[2] It is the patient work of generations of farmers that has made it possible. For example, from the same wild plant, we now have white and red cabbages, cauliflowers, kohlrabi, Brussels sprouts, leafy greens and many more.

The Traditional Approach

Knowing how to propagate plants was part of the basic knowledge that every gardener acquired during apprenticeship. Seeds were freely exchanged as common property. This ensured food security: the food production that fed the family and the community was ensured by the production of seeds on each farm.

Local seed production had other positive effects:

- Varieties were adapted to local soil and climate conditions. They became more resistant and productive over generations.
- Crops and crop pests co-evolved over centuries, allowing plants to become even more tolerant.
- Traditional varieties received little or no fertilisers, supplied only with manure and compost from the farm. In addition, the farmers of the past had limited means of irrigation – simple watering cans. This resulted in hardy plants that required little water and fertility.

The traditional system of plant propagation did not necessarily result in high-yielding plants, but it did promote food security, resilience and self-reliance, and of course it did not require any chemicals.

This ancestral system led to the creation of a very large number of local varieties, fully adapted to their terroir. Most have disappeared over the past century. According to the National Federation of Organic Agriculture, "75% of the genetic diversity of cultivated plants was lost between 1900 and 2000. 80% of the vegetable varieties cultivated fifty years ago have disappeared."[3]

▼ *At Bec Hellouin Farm, our aim is to grow only the tastiest vegetables by identifying hardy varieties that are well adapted to our soil and climate context.*

▲ *Runner bean seeds are particularly spectacular. During school visits, we encourage the children to open the long green pods and discover a plant version of Smarties sweets. What a surprise for them when the bright colours appear! Marvels like this make them take an interest in the plant world and capture their attention for the remainder of the farm visit.*

The Contemporary Approach

The first specialised seed producers appeared in Europe about 150 years ago. Today, the global seed market is dominated by seed and plant breeders – indeed, it is controlled by a dozen or so multinationals that also offer chemical fertilisers and pesticides. Seed production on the farm has almost disappeared. It is usually discouraged or in some cases penalised by legislation. Ancestral knowledge has been forgotten.

Official catalogues of seeds authorised for sale, such as the European catalogue, promote the standardisation of cultivated plants and the disappearance of old varieties because the cost of registering a variety is such that only specialised seed companies can afford it. This leads to the development of hybrid seeds that are not reproducible. In 1999, 73% of carrot varieties and 80% of tomato, pepper and Chinese cabbage varieties registered in the European catalogue were hybrids, and this percentage is constantly increasing.[4]

The shift from the traditional to a contemporary approach has resulted in a complete reversal of perspective, with far-reaching consequences. Modern varieties are produced in laboratories and research stations, under ultra-controlled and optimal conditions of temperature, fertility and irrigation, without any weeds, in order to obtain high-yielding plants. From then on, it is up to each farmer to try to reproduce the ideal conditions of those research stations on their farm, which makes them dependent on pesticides, herbicides, fertilisers, irrigation systems, and more. This dependence comes at a cost. Small farmers cannot always afford it, one of the factors driving the growth in farm size. In India, the switch from local varieties to GM varieties has led to the suicide of 100,000 small farmers in just a few years.

We have moved away from a system that produced hardy plants that were adapted to each locality, to a system where farmers must adapt to the seeds they buy from breeders. Yields are often higher, but farm independence is lost and biodiversity has diminished. Is such a system sustainable?

An approach to seeds based solely on the profit motive of specialised companies threatens our food security. In the event of a severe societal, economic or environmental crisis, seed production will no longer be secure. Nowadays, almost no one knows how to produce life-giving seeds locally (this applies more to vegetables than to grains, which can be resown as long as they are not GMO!).

Reproducing Your Own Plants

This shows how important it is to regain our autonomy. Seed collection has many benefits:

- o Maintain the specific and genetic diversity of varieties.
- o Perpetuate old or rare varieties.
- o Adapt varieties to your soil, climate and growing methods.
- o Benefit from varieties that are adapted or tolerant to diseases and pests (co-evolution between plants and pests).
- o Find new varieties.

However, seed production requires:

- o Time
- o Patience
- o Knowledge (especially in botany and genetics)
- o Skills
- o Rigour
- o Good planning, excellent follow-up; you have to be there at the right time
- o Some equipment and a dedicated room.

All market gardeners lack time, especially in the summer, which is why most of us buy our seeds. However, ownership of seed varieties for the benefit of commercial firms has recently led to a protest movement. An increasing number of people are getting involved in producing their own seeds, both amateur gardeners and farmers. Sharing networks are being organised. Reproducing plants is a civic and political act!

At Bec Hellouin, we have worked hard to establish the resilience of our farm, at all levels. But a few years ago, while discussing it as a team, we realised that in the event of a sudden collapse of society, we would be autonomous for everything, except seeds! Our aspiration to feed the local community in a crisis would collapse within months. So, seed production moved up the ladder of our priorities. We stopped buying F1 hybrid seeds whenever possible in favour of farmers' varieties (see p.18). Some members of our team have trained with our friend Philippe Desbrosses, the great pioneer of preserving old French varieties.

Some seeds are easy to obtain, like peas, beans, squash, tomatoes, chard, and arrowroot. Others are more difficult, like carrots, cabbage, and radishes. I would love to achieve complete autonomy. This fascinates and frightens me at the same time because, for a commercial farm like ours, the difficulties are real. I was very impressed by discussions with Grete, one of our trainees. She is a farmer from Holland who has been producing her own seeds for thirty years. She grows only a few varieties, well adapted to her terroir, with very versatile seeds, and only one leek variety that is suitable for all four seasons. She pointed out to me that a single plant usually produces thousands of seeds, and that only a few seedlings of each variety are needed to ensure production for the following year. This is great encouragement to move forward in this direction.

Generative Reproduction and Vegetative Reproduction

Reproducing plants by seed is not the only method of propagation available to us. There are two ways to reproduce plants.

Generative (Sexual) Reproduction: through Flowers and Seeds

This method of reproduction is based on the union of pollen (male genetic material) and an ovum (female genetic material), which produces a seed. The genome of each 'parent' is combined, so the seed is never completely identical to them. This allows for genetic mixing that favours adaptation to environmental conditions.

Vegetative (Asexual) Reproduction: Using a Part of the Plant

This mode of reproduction produces numerous seeds capable of going dormant and waiting for favourable conditions to germinate.

This mode of reproduction consists of taking a piece of just one plant. Various methods including cutting, air-layering, root, clump splinter, division of sucker, tuber, bulb, stolon, are all different ways to give birth to a new plant, genetically identical to the mother plant. The possibilities of mutation are reduced. As a result, the capacity to adapt is low.

In the next chapter we will discuss generative reproduction (seed production), before turning to vegetative reproduction.

"It was common practice until a few decades ago to grow and harvest local, native varieties that were inherited along with the farm, the garden and the fields. They were, so to speak, part of the family, perfectly adapted, thanks to generations of selection, to the growing conditions that the family practised and to their culinary tastes."

Andrea Heistinger[5]

2

SEED SAVING

The seed, the astonishing essence of life encapsulated in its tiny plant casing, the first link in the food chain, occupies a very special place in agriculture: Our cultivation practices, our dependence on chemical inputs, the quality of our food, and the perpetuation of our traditions depend to a large extent on the seed, its past interactions with man, its structure and genetic information, and its legal status.

Blanche Magarinos-Rey[6]

Being able to produce one's own seeds is also an act of resistance to a globalised system that privatises living things and degrades biodiversity.

To control the production of vegetable seeds and avoid degeneration (vegetables tend to revert to their original wild form), it is essential to have some knowledge of botany.

Flowers, the Reproductive Organs of Plants

All flowers are composed around a basic module, a peduncle ending in a receptacle bearing:

- Perianth
  - **Sepals**, collectively called the calyx.
  - **Petals**, collectively known as the corolla.
- Sexual appendages:
  - The **ovary** (future fruit), filled with one or more ovules (female parts, future seeds), topped by a **pistil** and ending with the **stigma**.
  - The **stamens** produce **pollen** (male parts).

Plants have adapted and evolved this basic module over time, so that there is now a great diversity of flower forms and floral organisations. This diversity has a direct influence on fertilisation and thus on seed formation.

As far as their mode of reproduction is concerned, flowers follow two patterns.

Bisexual (Hermaphrodite) Flower

The flower has both male and female reproductive organs. For example: tomato, cabbage, bean, pea, carrot, lettuce, onion flowers...

Unisexual Flower

The flower has reproductive organs of only one sex. It can be either a male or a female flower. For example: zucchini flowers.

The Different Types of Plants

Depending on the type of flowers they bear, plants can be classified into several categories.

The Plant Bears Bisexual (Hermaphrodite) Flowers

This is a hermaphrodite plant. This is the case for most species, e.g. tomatoes, but also for all Solanaceae, Liliaceae, Asteraceae, Brassicaceae, Lamiaceae and Apiaceae.

The Plant Bears Unisexual Flowers

Both Sexes are Found on the Same Plant

A **monoecious plant**: It bears both female and male flowers. For example: birch, alder, castor, squash, courgette, cucumber, pumpkin.

Both Sexes are Found on Separate Plants

A **dioecious plant**: The male and female reproductive organs are borne by different individuals. Some plants bear only female flowers (female plants), while others bear only male flowers (male plants). For example: willow, hops, bryony, kiwi, hemp.

◄ *A Dutch market gardener, Grete, gave us the first seeds of this leek variety, which we have been propagating on the farm ever since. It is a hardy and very flexible variety that can be grown all year round. This market gardener, who has been propagating her seeds for a long time, made me realise that you don't need to grow hundreds of varieties; it's better to identify a small number of 'sure values' adapted to your farm. Multiplication at home allows you to perfect their adaptation.*

▲ *Extraction of tomato seeds in the greenhouse. To separate seeds from the surrounding mucus, they are soaked in water for two days.*

Modes of Fertilisation and Risks of Cross-Fertilisation

Male reproductive cells (pollen) and female reproductive cells (eggs) can meet in different ways.

Allogamy and Autogamy

Cross-Pollinated Plants

'Allogamy' comes from the Greek allos, 'other', and gamos, 'marriage'.

In **cross-fertilisation**, gametes (pollen and ovule) are from different flowers. The ovule of one flower is fertilised by the pollen of another flower (the latter being positioned on another plant of the same species or, more rarely, on the same plant).

This situation is found in:

- **Dioecious plants**: obligatory allogamy (willow, hops, asparagus, kiwi, bryony).
- **Monoecious plants**: obligatory allogamy (oak, chestnut, hornbeam, squash, courgette, cucumber).
- Certain **hermaphroditic plants** (self-sterile flower or self-sterile plant): self-fertilisation is made impossible by various physiological or morphological processes (Liliaceae, Asteraceae except lettuce, Apiaceae, Brassicaceae, Lamiaceae, etc.).

In order to successfully reproduce cultivated plants, it is necessary to acquire some basic botany, a science that has been unjustly disparaged for decades but fascinates market gardeners!

For these plants, pollen is transported in different ways:

- By insects (bees, bumblebees, wasps, butterflies, flies, hoverflies, etc.): **entomophilous plants.**
- By the wind: **anemophilous plants.**
- By other vectors: birds, bats, small mammals, water...

For the production of cross-pollinated plant seeds, the distance between seed carriers of different varieties will therefore be directly correlated to the range of action of insects and wind. Otherwise there is a risk of cross-breeding and hybridisation between varieties.

Self-Pollinating Plants

'Self-pollination' comes from the Greek words *autos*, 'self', and *gamos*, 'marriage'. In the case of **direct fertilisation** or **self-fertilisation,** the gametes (pollen and ovum) are from the same flower.

This situation is found in certain **hermaphroditic plants** (self-fertilising flowers): each flower carries both the male gametes (pollen) and the female gametes (ovule). For example: oats, barley, wheat, sorghum, rapeseed, cotton, beans, flax, peas, beans, soya, Solanaceae (chilli, tobacco, tomato, aubergine, potato, pepper, etc.), lettuce.

The low level of genetic mixing results in faithful reproduction of the characteristics of the parental plant.

Strict autogamy is relatively rare, and there are often residual outcrosses (of the order of 5% in wheat, for example).

Who Crossbreeds with Whom?

The above information has many consequences for seed production. Unwanted crosses between different varieties should be avoided, as the new plants may lose the desired characteristics of both parents' varieties.

It is important to know when plants can cross. This information is provided by the botanical classification system.

Botanical Classification

In 1735, the botanist Carl Linneaus proposed a **binominal classification system** based on the anatomy of flowers: each plant is identified by a **genus** name and a **species** name.

The basic measure of this classification is the species. But the species is only one level of qualification among others:

phylum → class → order → family → genus → species → variety → sub-variety → race or cultivar

The Species Barrier

Only plants belonging to the same species can reproduce with each other and produce fertile offspring. This fact should be kept in mind when cross-breeding. In nature, this barrier is only crossed in exceptional cases.

Domestic carrots (*Daucus carota* subsp. *sativus*) and wild carrots (*Daucus carota* subsp. *carota*) belong to the same species and they can cross.

Cucumbers (*Cucumis sativa*) and melons (*Cucumis melo*) belong to different species; they cannot crossbreed.

Controlling Cross-Breeding

All the above botanical knowledge allows us to understand what each plant type is likely to cross with and its mode of reproduction.

In order to avoid unwanted crosses, there are several ways to isolate the seed carriers from each other:

- **Spatial isolation:** respect a minimum distance. It depends on the shape and vegetation of the land, the presence of natural (hedges) or artificial (buildings) obstacles, the number of seedlings, the floral diversity, the attractiveness of the flowers for insects, the wind direction, and more. More diverse environments, like permaculture farms, mean less pollen will be transported over a long distance.
- **Isolation by timing:** it is possible to stagger the growing cycles of two varieties that may cross – for example, growing one variety of lettuce in late spring and another in late summer. This is only possible for short-cycle crops and/or in regions with a favourable climate. Or, more simply, you can dedicate a season to each variety: make seeds of one lettuce variety one year and another the next.
- **Mechanical isolation:** seedlings can be grown in tunnels or under row cover. If they are pollinated by insects, introduce pollen-carrying insects like flies and bumblebees to ensure pollination, or provide manual pollination.
- In good conditions, such as those of a permaculture farm:
- **A wind-pollinated cross-pollinated variety** should be isolated 300-1,000 metres from another variety with which it is likely to cross. The distance should be increased if it is growing in the direction of prevailing winds.
- **An insect-pollinated cross-pollinated variety** should be at least 150 metres away from varieties that it is likely to cross with.

For self-pollinated plants, 2-10 metres should be maintained between varieties to minimise residual cross-pollination. For more details, see the vegetable factsheets in volume three.

Professional seed companies use larger distances.

Assessing the Risk of Cross-Breeding

In the excellent book, *The Manual of Seed Saving: Harvesting, storing, and sowing techniques for vegetables, herbs, and fruits,*[7] based on the experience of the Austrian Arche Noah and the Swiss association Pro Specie Rara, Andrea Heistinger suggests the following guidelines for determining the risk of cross-breeding.

Do several of the varieties grown for seed belong to the same botanical species?
No: no risk of cross-breeding.
Yes: potential risk of cross-breeding.

Are the plants self-pollinating or cross-pollinating?
Self-pollinating: no or limited risk of crossing.

▲ *Fennel seeds, on the stalk, in early autumn.*

▼ *Drying the best sunflower heads in preparation for the seed harvest.*

▲ *For vegetable growers, it is important to build up a relationship of trust with preferred seed suppliers. Their advice will help with choosing varieties to be grown. Feedback from growers can guide the work of the seed company. We particularly appreciate the Agrosemens team, for their ethics, their friendliness and the quality of their seeds and catalogue.*

▼ *New organic F1 varieties have interesting colours and properties. If we want to do without F1 seeds, it is essential to explore heritage varieties to identify old, hardy plants that can replace them advantageously. An exciting pursuit!*

Cross-pollinating: risk of crossing.

In the case of cross-pollination: is pollination done by insects or by wind?
By wind: isolation is needed, or protective tunnels.
By insects: isolation in the necessary space or use of isolation cages.

In the case of isolation cages: can plants pollinate themselves?
Yes: pollinating insects are not necessary.
No: insect pollinators are necessary.

A Word about Genetics

Before addressing the question of selection, it is advisable to know a founding principle of genetics. The 'father' of this science is a monk, Gregor Mendel (1822-1884), who discovered the basics of genetics by crossing peas with different phenotypic characteristics like yellow, green, smooth and wrinkled, and observing the evolution of these characteristics in the offspring.

Mendel deduced that each gene is present in two copies, or alleles, in the cell nucleus. These alleles can carry different information, either green or yellow in colour.

The information can be:

- **Dominant:** the presence of a single yellow allele, a heterozygous example, is sufficient to give the yellow colour to the pea;
- **Recessive:** the presence of a single green allele is not enough to give the pea a green colour; both alleles must carry the 'green colour' information for the pea to be green (**homozygosity**).

Codominance relationships exist: crossing a white flower with a red flower can, for example, produce a pink flower.

The Special Case of F1 Hybrid Seeds

A few genetic facts reveal how F1 hybrid seeds are produced.

First-generation hybrids or **F1s** are produced by crossing two **fixed or pure true varieties**, obtained after several successive self-fertilisations. They are homozygous: the two alleles of all their genes are identical (heterozygous: for each gene the allele of each parent variety is found).

The use of F1 hybrid seeds is authorised in organic farming and dominates the market for vegetable seeds intended for professional growers. They are very uniform, have high productivity levels and strong resistance to various diseases, which explains their rapid market growth.

However, F1 hybrids do not correspond to the natural cycles of life: they are generally obtained through biotechnologies, in seed company laboratories by fusion of protoplasts and other means. They cannot be perpetuated on farms because the characteristics of their parents are not duplicated. They have to be purchased every year. Hybrids cannot be adapted to local conditions.

The most beautiful hybrids of the second generation produce **F2 hybrids**. In theory, this work, repeated from generation to generation, makes it possible to stabilise the hybrid and obtain good homogeneity. I don't know of any growers who have tried; many hybrids are patented by their breeders.

Although they should not be confused with GMOs (genetically modified organisms), whose use is prohibited in organic farming, F1 hybrids, despite their advantages, do not fit into our vision of ecoculture and we avoid using them.

Selecting Seed Carriers

Sexual reproduction is a vector of heterogeneity: seeds from the same plant carry different genetic information.

Genetic variability ensures that a species can adapt to changes in its environment. The greater the variability, the more likely it is that a seed will carry a gene that adapts to its new environment or to an emerging disease to perpetuate the species.

If seeds are resown each year without selection, genetic variability will lead to the emergence of new traits. These can be either beneficial or harmful for agricultural production, as plants tend to revert to their wild form. Cabbage heads, for example, will be less dense and their leaves tougher; lettuce will be smaller. Characteristics of the original variety eventually disappear.

Selection and adaptation are natural phenomena. However, the criteria used by nature are generally not the same as those used by farmers.

Selection consists of choosing the plants that meet the criteria to be developed, and making them the seed carriers for the next generation. They should be marked with a label or a ribbon.

Selecting requires know-how, patience and a lot of time!

Selection Criteria

The first step is to define the objectives and criteria of the selection. If you harvest your own seeds, it is up to you to choose, and this is where the process becomes exciting.

Here are the main criteria for selecting seedlings:

- **Good germination:** quick emergence and good seedling quality.
- **Regular growth:**
  - Regularity and length of cycle before harvest.
  - Growth speed: early – grows early in the year or has a short cycle, late – grows late in the year.
- **Conformity to the morphological characteristics of the variety** including shape, colour, foliage, height, and size.
- **A tolerant or resistant plant:**
  - Resistance to climatic hazards such as heatwave, drought, frost, and humidity: climate is a valuable aid in the selection process, so keep seeds from plants that have survived a difficult climatic episode.
  - Resistance to bolting for leafy vegetables, root vegetables and leafy herbs.
  - Tolerance to diseases and pests. Genetic variability is an important consideration to attain these traits.
- **Flowering and/or fruiting quality:**
  - Size, shape, colour of fruits and/or roots.
  - Flower morphology or colour.
- **Harvest quality:**
  - Productivity and yield: total quantity, regularity, size.
  - Shape.
  - Colour.
- **Taste quality**
- **Nutritional quality**
- **Conservation.**

The criteria of the agro-industry are: suitability for mechanisation, tolerance for transport and shelf life, synchronised maturity, size, resistance to freezing or cooking... This is not the case for microfarms!

Be aware that it's all a question of balance. You have to play with different factors; pushing one too far, like conservation or yield, means degrading others like flavour and disease resistance.

Different Types of Selection

There are different ways of achieving the objectives that we have set for ourselves.

Nasturtium seeds

Conservation Breeding

Objective: to maintain genetic diversity and prevent extinction.

The maximum number of characteristics specific to a variety must be preserved. A description of these characteristics can be found in seed company catalogues or at the International Union for the Protection of New Varieties of Plants (UPOV).[8]

Ideally, the most and least beautiful seedlings should be selected at random, so as to have all the distinctive characteristics of the variety.

Broccoli seeds

Positive Mass Selection

Objective: to improve a variety.

Only the most beautiful plants are selected to be seed carriers.

Only seeds from plants that best meet targeted objectives are kept.

Negative Mass Selection for Purification

Objective: varietal purification.

Eliminate seeds that least meet targeted objectives. Prevent these plants from flowering.

Spinach seeds

Hybridisation and Natural Mutation

Objective: to create a new variety.

Natural mutation: spontaneous mutation occurring in a population. The breeder sows a large number of plants and looks for a possible interesting mutation, for example a different colour.

Hybridisation: crossing two varieties. The crossing of two pure varieties results in a first-generation hybrid called an F1 hybrid (see previous page).

How to Stabilise the Characteristics of New Seeds

If a variety is developed or a new one is created, it must be stabilised, while maintaining its genetic diversity.

Bean seeds

For Self-Pollinating Plants

Genetic diversity is weak and cross-breeding is rare. Sensitivity to inbreeding depression is low.

Corn seeds

Parsnip seeds

Pea seeds

Stabilisation is relatively quick to achieve. At each generation, select seeds that correspond to the desired criteria and eliminate those that do not.

A few seed producing plants are sufficient to maintain genetic diversity.

For Cross-Pollinated Plants

Genetic mixing is important (cross-fertilisation). Sensitivity to inbreeding depression is high.

Genetic stabilisation takes longer to achieve (8-10 generations).

Maintain **several dozen seed producers** in order to ensure sufficient genetic mixing and to avoid defects (the equivalent of inbreeding problems in the animal kingdom).

Questions to Ask before Producing Seeds of a Particular Variety

On the following page is a seed production sheet based on the work of Andrea Heistinger.[9] We suggest that you record all your tests in a notebook or binder. You can make a sample sheet for each variety with the following questions. Then write down all your observations, year by year. This way you can follow the evolution of your variety.

Seed Production Step by Step[10]

Cultivation of Rootstocks

Growing seed producing plants is not very different from growing other vegetables, it is simply a matter of letting nature take its course and completing the plant's cycle. However, to obtain a good harvest of vigorous seeds, it is important to ensure that this cycle is carried out in the best possible conditions.

The procedure is the same as vegetable cultivation, consisting of soil preparation, planting, amendment and fertilisation, protection against diseases and pests.

Spacing: plan for wider spacing between plants than what is given for conventional crops, or remove plants between seed producers. Flowering and fruiting plants need more heat and sunlight.

Watering: the seed producing plant can be stressed so that it develops less need for water resources and a stronger root system. However, care must be taken to avoid water stress at flowering, as this could lead to a deterioration in the quality of the flowers and seeds. Stop watering and protect from rain during the final phase of seed maturation.

Fertilisation: take care not to damage seed carriers. Handling care is the same as for sale or consumption.

Provide **a staking or trellising system.** The flowering tops must not come into contact with soil moisture.

Prune flowering stems to increase the number of branches, flowers and seeds; this technique is particularly interesting for the Asteraceae, Brassicaceae, Chenopodiaceae, and Lamiaceae. It is counterproductive for Alliaceae.

Keep more seed producers than necessary and then sort them as you go along. A large number of seed producers promotes genetic diversity.

Isolate the seed producers, as discussed on p.18, to avoid hybridisation.

When seed producing plants are isolated under cover and fertilisation is no longer done by wind, insects or birds, **manual fertilisation** may be necessary. This is easy to do for unisexual flowers like Cucurbitaceae, but much more complex in the case of hermaphroditic flowers as the stamens are removed from the female flowers.

Beware of greedy birds and other pests. Use traps, nets, repellents, and scarecrows, if necessary.

The Unique Case of Biennial Plants

Plant cycles are determined by their adaptation to local climate and resistance to cold: they are annuals, biennials or perennials (for more details, see volume one).

For seed production, biennials are difficult. The plants selected for seed production must overwinter in the best possible conditions, protected from bad weather, pest attacks and diseases that thrive in cold and humidity. Tubers are most often harvested at the end of their vegetative cycle, at the end of the first year, and stored in a silo, cellar or in any cool, dark place for preservation. They can also be transplanted into a cold, frost-free greenhouse.

In the spring of the second year, biennials are gradually acclimatised to light and outdoor temperatures before being transplanted. They should then be watered well to facilitate their recovery.

SEED PRODUCTION SHEET

Species:	Variety:
Origin of the seed:	

PRELIMINARY QUESTIONNAIRE

Is the reproduction of the variety:		Generative (by flowers and seeds)?		Vegetative (by part of the plant)?
With which other vegetables can it be crossed?				
Is the variety:		Self-pollinating (direct fertilisation or self-pollination)?		Cross-pollinating (cross-pollination)?
In case of cross-pollination, is the variety pollinated?		By insects?		By wind?
Is isolation necessary?		Yes		No
If so, what form of isolation should be chosen?				
Depending on the cycle of the variety and the weather in my area, when is the best time to sow seeds?				
How many carriers are needed?				
Is the variety to be propagated:		An annual plant?		A biennial plant?
If it is a biennial, what should I do for wintering?				
What are my selection criteria?				
What are the possible diseases and pests?				
What are the signs of seed maturity?				
What is the best time to harvest?				

FIRST MULTIPLICATION

Date of sowing:	
Number of seed carriers:	
Characteristics of the seed carriers:	
Date of seed harvest:	
Quantity of seed harvested:	
Method of storage:	
Observations on the next generation:	
Remarks:	

SECOND MULTIPLICATION

Date of sowing:	
Number of seed carriers:	
Characteristics of the seed carriers:	
Date of seed harvesting:	
Quantity of seed harvested:	
Method of storage:	
Observations on the next generation:	
Remarks:	

Pumpkin seeds

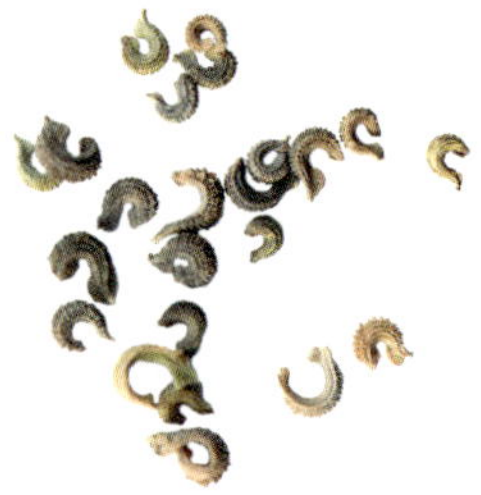

Marigold seeds

Tomato seeds

Sunflower seeds

Harvesting the Seeds

When to Harvest?

Harvesting is done when the seeds are visible, dry, hard and ready to fall, and when the floral terminals, no longer fed by the stems, dry out and turn brown. These indications are given in the vegetable charts in volume two.

Seeds can be harvested after watering: the residual moisture prevents seed loss (retained internally) without becoming too wet. If there is no risk of dissemination, harvest in full sun in the middle of the day.

Always choose seeds from several plants. This is essential to maintain sufficient genetic variability.

How to Harvest?

There are different ways depending on the plant:

- **Cutting the whole plant:** lettuce, peas, lamb's lettuce, beetroot.
- **Cutting off the heads:** poppies, sunflowers, onions, carrots.
- **Cut the stems:** borage, comfrey, cabbage, turnip, rocket.
- **Collecting up from the ground:** nasturtium, tetragonia (New Zealand spinach).
- **Picking fruit:**
  - When overripe: aubergine, cucumber, courgette.
  - When ripe: tomato, pepper, squash.
  - As you go along: dandelion, salsify, scorzonera, cosmos.

► *This chard seed producer is supported by stakes so that it does not touch the ground and succumb to rot. A single stalk can carry a considerable number of seeds!*

Use suitable containers: sheet, bucket, bin, paper bags, racks, for example. Beware of plastic sheeting – it retains moisture. Avoid putting crops in contact with the soil.

Plan several passes for plants whose seeds mature at different times (Apiaceae, Brassicaceae, etc.).

Be sure to label the plants that are harvested.

Seed Extraction

Once the seeds have been collected, they must be separated from their pods. There are two possible ways to do this.

Dry Processing: Threshing (Dried Fruit)

Dry your seeds in a dedicated place, on a sheet in a well-ventilated, shady, dry attic, protected from rodents and granivorous birds. Leave them for the prescribed time, one week for lamb's lettuce and one month for leeks, for example.

For threshing, let the dew evaporate first to keep seeds dry. Find something to thresh on – a simple plastic sheet will do. Lay out the plants to be threshed on the sheet.

When the weather is warm and the seeds are dry, carry out the threshing:

- Some seeds can be detached without effort: tetragonia, camomile, lamb's lettuce, poppy, snapdragon.
- **Threshing in a bag:** onion, leek, chicory, rocket...
- **Threshing in a container:** lettuce and similar crops.
- **Threshing in a sieve:** oregano and the like, in small batches.
- **Threshing by shelling:** shelling of pods like peas, beans, and chickpeas.
- **Threshing by rolling:** includes bean, beetroot, turnip, radish, chicory, pea, bean, and cabbage. On a soft surface like a mown meadow, covered with a sheet, use feet or a lawn roller.

Make two passes at two-hour intervals, aerating in between with a hay rake.

Wet Treatment

Used for fleshy and watery fruits.

Fleshy fruits: pumpkin, courgette, aubergine, lemon, etc. Allow the fruit to ripen as long as possible after harvesting before removing the seeds. A simple cleaning is then sufficient.

Watery fruits: tomato, cucumber, gherkin, and more. Harvest and allow to ripen in the shade. Then remove the seeds and the pulp

◄ *The flower heads of mature leeks are dried in the barn over a collection cloth.*

that surrounds them, soak them in water for liquid fermentation for about two days at 20°C, then wash several times before filtering.

To separate the good seeds from the infertile or poor quality seeds, which are lighter, they can be immersed in water: only the good seeds will sink. Collect them, then drain and dry them thoroughly.

Cleaning and Sorting Seeds

After seeds have been threshed using the dry treatment, they remain mixed with impurities like pieces of stems or leaves. To separate them, in the context of a garden or a farm, simple and traditional procedures are used:

- With **a set of sieves** numbered from two, the widest mesh, to 40, the finest mesh – the sieve number corresponds to the number of openings per square centimetre.
- By **hand** – winnowing and blowing.
- With the **wind** or a fan.
- With **the force of gravity**.
- With **centrifugal force** or a honeycomb sorter to separate the seeds.

Finish sorting by hand to eliminate the last impurities and seeds that are too small, of poor quality or diseased. Select seeds that are similar to the original seeds. They should be dry and healthy.

Drying Seeds

Seeds must be dry in order to ensure good conservation. Drying is done in a warm and dry atmosphere of not more than 40°C, ideally 30°C. Air circulation is more important than temperature.

▲ *A set of sieves of different mesh sizes is used to separate seeds from impurities.*

▲ *Orach seeds are particularly easy to harvest. We most often let this plant reseed itself in the garden.*

Drying should be as fast as possible. The aim is to achieve 6-12% residual moisture. Testing for dryness can be done with a fingernail or tooth, like our ancestors did. The seed must be brittle and not feel soft. Be careful; a seed that is too dry or cracked may be damaged. Be sure to protect seeds from greedy birds with screens and from strong gusts of wind - weight the fabric on which the seeds are spread.

Stir the seeds several times a day, as often as possible.

Drying can be done:

- **In the wind,** on racks or in coffee filters hung on wires. Put only a small quantity of seeds in each filter and identify them well.
- In a dry, well-ventilated **loft.**
- **By forced drying:**
  - Electric dryers or dehydrators.
  - Solar dryers.
  - Fan dryers or other sources of hot, dry air.
  - Drying in a domestic oven.

Storage of Seeds

Good storage conditions greatly improve the germination time of seeds and their germinative power. This is achieved by choosing suitable containers that protect seeds from moisture and by a cool, dry room that is protected from rapid and drastic temperature changes.

Containers: small cloth bags, sachets, paper envelopes, boxes... Avoid airtight containers (plastic bags, glass jars): seeds, even in dormancy, need oxygen. However, non-watertight containers do not protect against humidity, so the room should be dry.

Properly label: species, variety, date and place of harvest, and any other information deemed necessary, like characteristics and variety.

▲ *Small paper bags are well suited for saving seeds. A simple folding process can be done.*

Use wooden containers: for buffering and insulating capacity, but do not use chipboard or plywood containing formaldehydes. These are germicidal products.

Check batches regularly.

Protect from pests: voles, mice and rats, flour moths, and bruchids, for example.

Protect from humidity and heat:

- Ideally in an airy loft.
- The lower the temperature and humidity, the longer the life of the seeds. A refrigerated room with a temperature of 1-5°C increases the shelf life.
- Target a humidity level of 40-60%, ideally below 40%.
- Target a temperature of 15-16°C, ideally below 12°C.

Seed Treatment

Some seeds may carry diseases like fungi, viruses, and bacteria or pests like bruchid beetle, for example.

Monitor the crop: remove plants with disease symptoms.

Neutralise undesirable organisms:

- Freeze at -18°C for at least 24 hours or even a week.
  - Protect seed packets in an airtight container so that there is no condensation from the freezer. Allow to come to room temperature in the shade.
  - Seeds put in the freezer must be perfectly dry, otherwise they risk being damaged.
- Heat therapy using hot water at 50°C for 30 minutes, then dry.
- Water vapour.
- Soaking in a vinegar or essential oil solution.

Germination Tests

It can be useful to check the quality of a batch of seeds with a germination test, especially if they are old. A low germination rate indicates a loss of vitality in the seed set, which will affect production.

Here are the steps to follow and the factors to consider:

- Test a batch of 50-100 seeds.
- Remember, to germinate, seeds need water, air, heat and light.
- Use a moist medium: blotting paper or cotton wool, sand, potting soil, for example.

- Carefully selected plants that have been isolated from their wild ancestor germinate faster and easier.
- Germination rate and time are averages; sometimes seeds do not germinate in vitro but do very well when sown directly in the ground during the right season.
- Generally, a good germination rate and rapid emergence go hand in hand.
- Sometimes it is necessary to break dormancy, especially for seeds of plants from temperate regions that are still close to their wild type, with flowers for example. Keep in the freezer for one to two weeks.
- Techniques to facilitate germination: stratification, in cold temperature or with thermal alternation; scarification and peeling away the seed coat to allow water and air to penetrate; soaking, for example with chickpeas and squash. Pre-germination for small seeds.
- Differentiate between germination capacity and germination vigour, resistance and vitality of plants after germination. A seed may have retained its germination capacity but lost its vigour.

◀ *It's best to use natural containers for seed storage. Old trade furniture with lots of drawers is great for seed storage cabinets.*

Different types of seeds

To help navigate the maze of names used to describe seeds, here is a non-exhaustive list of the most common.

Terroir varieties: varieties selected from the same terroir, perpetuated from generation to generation. They have adapted to particular soil and climate contexts. A terroir = a type of soil + a type of climate. Examples of terroir with a cultural connection: Carentan leek, Milan cabbage, cantaloupe melon, and Tarbais bean. These are also referred to as heirloom, farmers' varieties. They are in the public domain, unrestricted by intellectual property rights.

Standard or certified seeds: improved commercial varieties. The breeder may have a plant variety protection patent.

Farm-saved seed: seeds produced on the farm from certified seed.

True breeding: varieties bred after several successive generations become homozygous.

Hybrid varieties: varieties resulting from the crossing of two pure varieties are heterozygous.

Treated seeds: seeds coated with a treatment product, either a fungicide or an insecticide.

Graded seeds: their size is homogeneous, which is necessary when using a precision sowing machine.

Pelleted seeds: graded and treated, generally with a fungicide to prevent damping-off.

Coated seeds: with clay, peat or gypsum, sometimes combined with a plant protection product. Germination facilitated.

Pre-sprouted seeds: interesting for seeds with long emergence like celery, but their cost is higher due to labour time.

Seed tape (for seedlings in rows) or seed mats (for seedlings in nurseries): intended for amateur gardeners.

▲ *There is something deeply satisfying about saving your seed crops! It builds self-confidence and inner security in our ability to act in concert with nature.*

◄ *In the greenhouse workshop, Louise and Éléonore repot berry cuttings for planting in new areas of the farm, or for selling.*

3

VEGETATIVE PROPAGATION: CUTTINGS, TRANSPLANTS, DIVISIONS, LAYERS

Grafting, grafting, what did he need to graft?...

Dissatisfied with the coherence of a universe of which he was nevertheless the attentive witness, he only aspired to distinguish himself from it; and to do so, grafting in hand, he was going to leave his mark!

He was obliged to note that he was grafting like a madman at all times and in all places. He could not meet a wild species without immediately wondering how he was going to make it productive. But only by grafting could the rose be harvested from the wild rose!

Maurice Chaudière[11]

One day when I was walking in the Amazon with a group of Amerindian teenagers, I was surprised to observe that when they picked an edible plant to nibble on as they went along, they quickly replanted a cutting at the edge of the path, without thinking about it any further. By doing this for generations, the last hunter-gatherer peoples of the Amazon created edible forest corridors by concentrating useful plants along their trails. I am fascinated by this way of facilitating survival, by gently managing the natural environment.

If I had told my young native companions that they were practising a form of vegetative propagation, they would have been surprised. Planting a branch in the ground is like taking a cutting, it's as simple as that. If it grows back, i.e. if it puts down roots, the result will be a new plant, genetically identical to the one from which that piece of branch came.

Vegetative propagation is a simpler way of obtaining new plants. Moreover, some plants that do not produce seeds can only be propagated in this way. It can be done in various ways. We will briefly mention them here, as the subject is already well covered in existing literature.[12]

▲ *Already in the Middle Ages, peasant farmers mastered grafting techniques perfectly.*

Cuttings

The plant world has such plasticity that some cells can 'dedifferentiate', i.e. return to a primitive, embryonic state, and thus acquire new developmental potential that can lead to a new plant. A leaf cell can become a trunk cell, or give rise to roots.

When a branch, a stem, sometimes even a leaf is cut off and put back into the ground, a new plant can result. Most plants can be grown from cuttings, especially perennials: trees, shrubs, small fruits, herbs, flowers...

The cutting must be taken from a healthy and vigorous plant that is well adapted to its soil.

Germination Tests

Herbaceous cuttings are taken from non-woody plants, often at the end of the summer. We use them to propagate perennial plants.

Green cuttings, or softwood cuttings, consist of taking the year's branches when they are still young, before they start to harden. This type of cutting is taken between May and July. It is ideal for shrubs.

August cuttings are taken from the current year's stems, at the end of summer, when they are at the semi-woody stage, i.e. in the process of hardening and turning brown. This is the type of cutting we use most often for shrubs and bushes in our forest gardens or hedges. We prefer it to green cuttings because at this stage, the shrubs have generally given their fruits. It is used, for example, on blackcurrants, currants, gooseberries, chokeberries, saskatoons, mulberries, thornless blackberries, goji berries, elderberries, Damascus roses, willows, olive trees and vines. It is also suitable for lavender, sage and rosemary.

Node cuttings consist of taking a small piece of the node along with the cutting in the spring.

Leaf cuttings are a simple way of propagating succulents. This is the easiest way to propagate comfrey, that precious plant whose seed set is random. All you have to do is take some leaves and stick them in the ground where you want them to grow again. Of the plants on the farm, only comfrey and sedum can be cut in this way.

▼ *Propagating aromatic and medicinal plants in our large greenhouse, by cuttings or seedlings depending on the species.*

Root cuttings are used for some flowering plants and are also an excellent way to propagate comfrey. The practice is simple: just cut a root into pieces and bury it. In general, fleshy-rooted perennials and suckering shrubs are well suited to this method of propagation.

Heel cuttings are taken from woody plants. They are formed from a lateral branch taken by gently pulling on it or cutting it with shears, leaving a small piece of the main branch at its base. This type of cutting is done in spring and late summer. We hardly ever use this technique, as our needs are met with other cuttings.

Cuttings grown under cover. This is a technique that can be used with most of the cuttings described above. It adds a certain amount of security to the root system, but it is by no means a requirement. Greenhouse conditions maintain heat and humidity around the cutting, the aim being to encourage faster rooting. To do this, the cutting is covered, more or less hermetically, with a transparent material to generate heat and humidity: a hoop of plastic or a glass bell; cut-up 1 litre water bottles can also be used; some people just use plastic bags. The cuttings should be aired out regularly to avoid excessive condensation. This could cause leaching and lead to rot and mould. Herbaceous cuttings are particularly susceptible to this phenomenon. Covered cuttings should not be exposed to the sun for long periods of time: the temperature inside the confined space would quickly become excessive and burn the plant.

Cuttings in water is probably the easiest way to take cuttings of non-woody plants such as herbs, but avoid chlorinated tap water. Children are fascinated by the appearance of new roots! Once roots are visible, you can simply replant the cutting in soil.

Willows and poplars are easily propagated as cuttings, particularly for pollarded trees. Use a beautiful straight branch that is over one year old, 2-3 metres long, and push it into the ground 70-100cm. We have noticed that willow is so vigorous that cuttings are successful all year long.

How to Make Cuttings

Once clipped, the cutting is prepared:

- Leaves are removed from the lower third of its length.
- The aerial part is reduced to 5-15cm, depending on the species, and the leaves

◀ *To succeed with grafts and cuttings, timing and choice of healthy plant material is essential.*

are partly cut off in order to limit water loss through transpiration.

The cutting is then planted either in a dedicated nursery, or in its final location. In this case, the maintenance required is minimal. Weed well to avoid too much competition from grass in the first year. Watering will depend on soil conditions.

It can also be replanted in a pot. The substrate can be composed in different ways. Ours is made up of one-third compost (ideally woody), one-third sand and one-third soil. We mix the mixture in a wheelbarrow before filling the pots. We usually use repotting for fragile plants or those that suffer from a harsh winter. The pots are stored in a sheltered place like a greenhouse; care must be taken not to let the substrate dry out. The cutting and its root ball can be transplanted the following spring, as soon as the risk of frost is over.

Whichever option you choose, make sure that your little cutting does not suffer from stresses like too much or too little water, excess heat, cold or humidity.

For all cutting techniques, new roots appear within a few weeks.

Rooting Hormones

A rooting hormone is a substance that helps a cutting withstand stress by preventing it from closing, or healing the cut part. This makes it easier for new roots to grow. It can be purchased and you can also produce it yourself from willow, which contains salicylic acid, an ingredient in acetylsalicylic acid, commonly referred to as aspirin.

To make this hormone, soak a few 5cm sections of willow branches to extract the active substance, in a bucket of non-chlorinated water. In three weeks the solution becomes a gel, with an aspirin odour, that you can coat your cuttings with before putting them in the ground. At Bec Hellouin Farm, the vigour and resistance of the plants we take cuttings from means that we do not actually need to use this cutting hormone. But it can facilitate the recovery of certain difficult species.

Clump Division

Dividing a well-established perennial clump is easy. In wet weather, simply dig up the clump and cut it into pieces of about 10cm by 10cm with a spade, fork or knife, from the periphery of the plant. It is not necessary to uproot the whole plant. This operation is best carried out when the plant is not growing, in autumn before the cold season (some plants are better in spring). Dividing also has the effect of restoring the vigour of the plant, the central part of which tends to wither when it becomes too dense. Water well when planting the divided parts.

Division by Stolons

Strawberry plants produce stolons, new plants connected to the mother plant by a long stem, a kind of 'umbilical cord' that feeds them until they produce new roots. The seedlings from

▲ *Grafting makes it possible to keep many varieties of fruit trees intact, from generation to generation.*

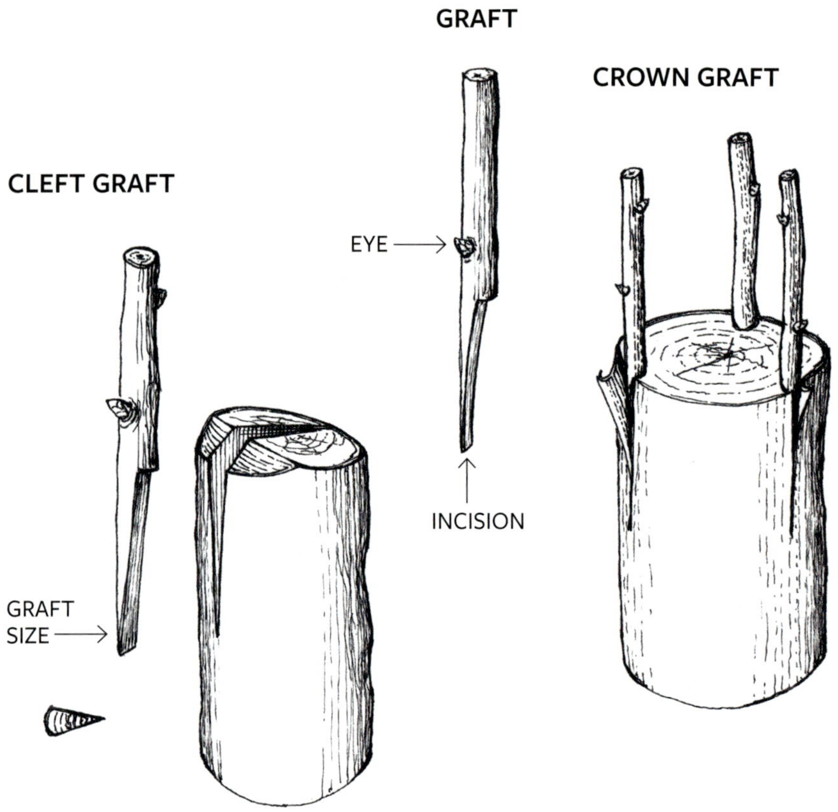

the best plants can be transplanted into a potting mix when their roots emerge. Keep shaded until they are well established.

Division by Separating Suckers

Suckers are young shoots that grow from the roots of a plant, along the periphery. They are removed by gently peeling them from the main root system with a few roots. They are then replanted in a nursery. Many trees, shrubs and perennial flowers can be propagated in this way, including walnut, poplar, and even cherry and apple trees. If these come from grafted trees, they will have the characteristics of the rootstock, not the scion.

▸ *To obtain large sizes, such as this nut, it is essential to choose grafted varieties rather than wild ones.*

Division of Bulbs, Tubers and Rhizomes

Many plants can be propagated by separating their underground organs from each other: bulbous plants, Chinese artichoke, jerusalem artichokes, irises, dahlias, to name a few. These divisions are carried out during the vegetative rest period.

Layering

The branches of many woody or climbing species like cucumber, tomato, kiwi, blackcurrant, and currant root when in contact with soil. This property can be used to make clones. To do this, a branch is bent and buried with a wire hoop, stone, or log, either in the ground or in a pot. Once the branch is well rooted, it can be separated from the parent plant with shears and replanted elsewhere.

For some plants, you will not have to make any effort because they lay down roots of their own accord. Blackcurrant, whose low branches bend to the ground, lays down roots without having to force this contact. Natural layering of a single branch can result in several new bushes.

Layering can also be used to give a plant a new lease on life. We have experienced this with cucumbers. One year, the first crop of cucumbers was weakened by late frosts, and aphids feasted on them. The tops were still green and productive. We had to remove the damaged leaves from the bottom to midway to save them since the second crop was not fruiting yet. We layered each of the weakened vines and prolonged production for more than a month.

When it is difficult or impossible to bring a branch into contact with the ground, it is possible to make the ground come to the branch. This is called aerial layering. A plastic bag filled with substrate, like that used for cuttings, is attached to the branch and remains there for several weeks.

When the layer is separated from the mother plant, its root system is sufficiently developed for planting in its final location. Planting should preferably be done in spring.

Grafting

Grafting two plants of the same botanical family combines the qualities of each variety. Scions are taken from a desired variety and attached to a rootstock, a 'support plant' that only has roots and a stem.

Grafting[13] is used to propagate certain species of woody plants, especially fruit trees, when seedlings of pips or stones do not give satisfactory results, when their growth is very slow or when they are poorly adapted to certain soil types.

This operation is rarely carried out by amateurs, as it requires more knowledge and attention to detail than cuttings. While it can be done by anyone, training from an experienced professional is a good idea. The following information is only intended to give general guidelines.

Rootstock

The choice of rootstock is important because it gives plants various characteristics like vigorous or weak bearing, good adaptation to particular soils like heavy and chalky, resistance to certain diseases, speed of fruiting, and more.

Different varieties

Grafting is done on different types of fruit trees and shrubs. It is very easy to take a few grafts from an interesting variety, for example from an old local variety that is no longer offered for sale but that you discovered in a neighbour's orchard. You can also find interesting grafts at tree fairs or from conservatory orchards or certain regional nature parks.

How to Graft

The grafting operation takes place in three stages - preparation of the rootstock, preparation of the grafts, and then fixing the grafts to the rootstock.

Grafting is usually done in March-April, when sap flow is strong. There are different types of grafting.

The principle consists of cutting the bark of the graft holder and inserting a branch bearing one or more eyelets.

Care must be taken to ensure that the plant tissues are correctly positioned and that the cambiums of the rootstock and the scion are facing each other, so as to ensure the passage of sap between the two pieces. They are joined together by a ligature and grafting putty or wax.

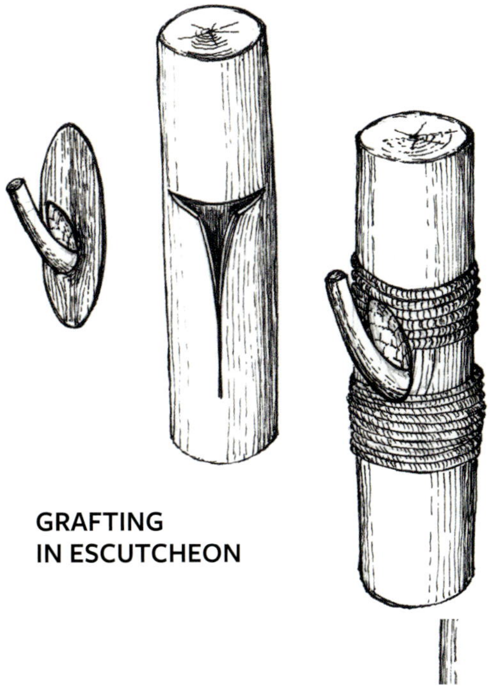

GRAFTING IN ESCUTCHEON

◄ *Grafting knife and folding pruning saws.*

Pulverised charcoal has a sanitising effect, and can be applied to cut parts of plants to prevent disease.

The techniques for propagating plants discussed in the previous chapters allow anyone to create a microfarm in a few years on a small budget. This process does require initial training and takes more time; however, the advantage is that the market gardener gets to know plants intimately, and establish a strong relationship, as well as reducing costs.

▼ *This young plum tree broke in a storm. Of the four crown grafts that were done, only one has recovered. It is not too late to try again!*

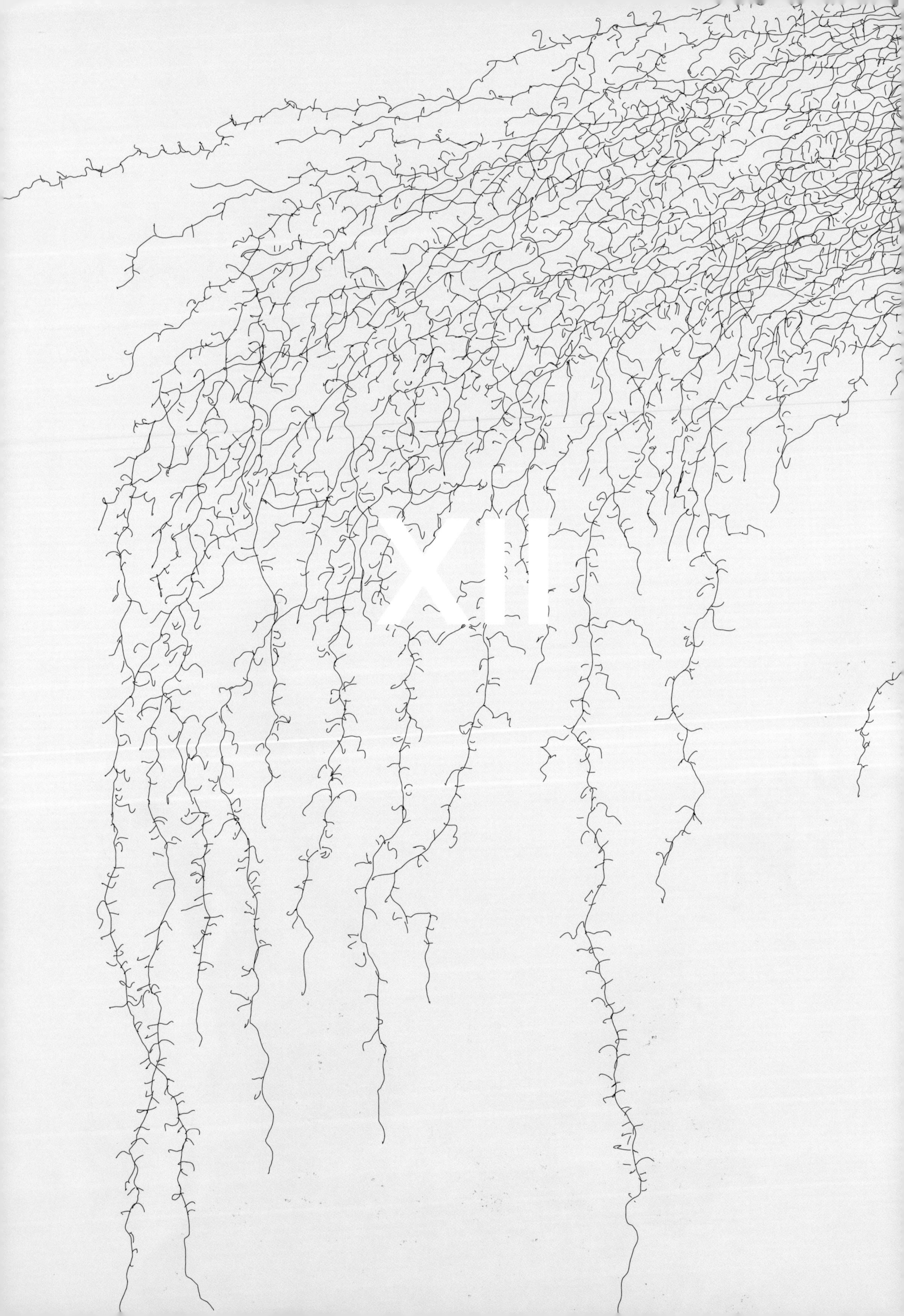

XII

GARDEN HEALTH AND BIODIVERSITY

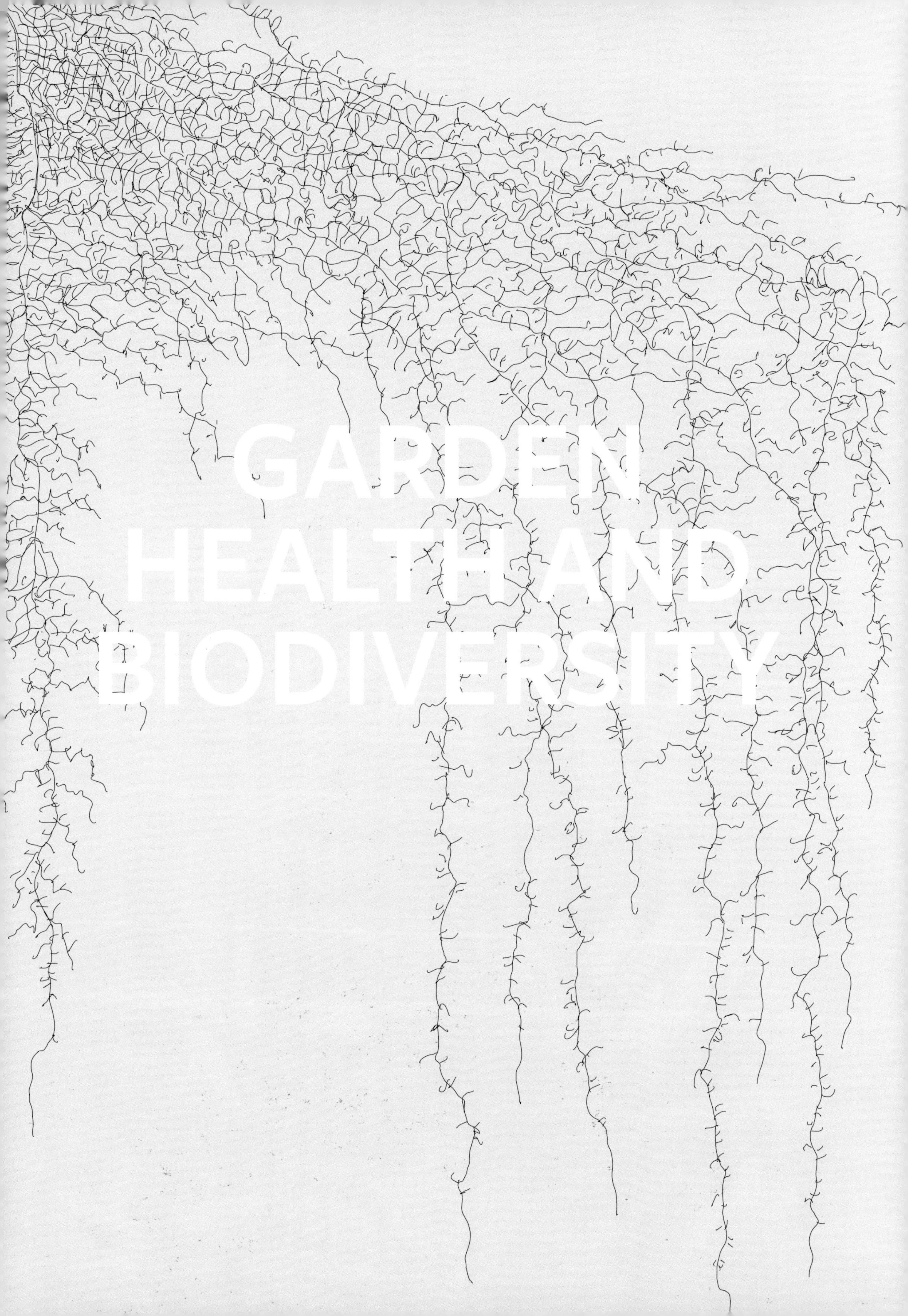

◀ *To most neophytes, all bees live in hives and make honey. Yet honeybees,* Apis mellifera, *like this specimen foraging on a poppy flower, form only a tiny fraction of bees and are less prolific pollinators than their wild cousins. The mason, mining and bumblebee, shown below, are among the wild bee types present on our farm.*

4

HEALTH AND VITALITY IN THE GARDEN

I know an organic farmer who grows 50 hectares without an ounce of chemicals and whose production per hectare – for all crops – exceeds national averages, and who says he can show you all the wheat diseases present, but none of them developed enough to affect his production in the slightest.

John Seymour[14]

Protecting crops from pests and diseases is a major concern for farmers – for some, an object of torment. I must confess that at Bec Hellouin we have devoted little interest to this subject, apart from the management of slugs and voles, and downy mildew in tomatoes: simply because it is rarely a concern. To be honest, I don't know much about small garden pests, but working on the topic for this manual has allowed me to put our experience into perspective and better understand how natural regulation occurs on our farm. In fact, many of our actions contribute, directly or indirectly, to its good health.

Mason bee

Mining bee

Bumblebee

Encouraging Encounters

The approaches described in this manual contribute in one way or another to garden health, including soil care, plant nutrition, the complexity of the environment, crop combinations and more. Thanks to all these measures, we have been able to observe how a balance between pests and beneficial allies of the gardener – birds, insects, amphibians, hedgehogs and reptiles – that feed on pests, has gradually been established over the years. This has been particularly noticeable for slugs, who thrive in raised bed cultivation and mulch. We have actively sought a balance. Our garden is not free of disease and pests, but the level of damage is rarely of concern and rarely warrants intervention on our part.

A few years ago, I was invited to carry out an interesting consultancy project on Réunion Island. I visited a series of agricultural installations that were facing various difficulties. Their managers wanted to develop them using permaculture as a model. When I raised the possibility of switching to organic farming with the technicians, a strong, almost visceral resistance arose each time, "It's impossible here, the pest pressure is too strong!" It is true that in the tropics, the year-round hot climate does not offer a break from pests and disease development; however, looking at these monocultures growing on bare soil, in a highly artificial environment, I could not help but think that by renaturing these facilities, a positive result would arise. "How did your parents do it? They didn't use pesticides," I asked the technicians. They just nodded and said that this was no longer possible. They said they were surrounded by swarms of hungry predators against which the fight is intense and relentless.

We looked for farms that exemplified another way forward and were delighted to meet satisfied farmers, including Max Fontaine, an agricultural engineer turned organic market gardener and permaculturist. For Max, going organic had been the solution, not the problem! His market gardens were the result of a long process of reflection and research, and included a large number of species, many of them local varieties. The diversity of crops flourished without the use of chemicals. During a visit to CIRAD (Centre de Coopération Internationale en Recherche Agronomique pour le Développement), we

▲ *This is Max Fontaine's garden on Réunion Island. He has proven that crops grown in as natural an environment as possible are less likely to be attacked by diseases and pests than denatured plants.*

met a friendly and helpful agroecologist, Jean-Philippe Deguigne, without knowing that he had become an international expert on natural pest control. When I asked Jean-Philippe if it was possible, in the context of a tropical island, to do without plant protection products, he laughed and said: "Of course it is possible: it is these products that make the plants sick!" He then explained to me that his research, carried out over more than twenty years, had convinced him of the effectiveness of a natural approach. For example, Jean-Philippe has developed a simple method for combating the vegetable fly, the main scourge in Réunion. Diseased fruit and vegetables are collected in order to prevent infestation of the plots, and placed under a wire mesh bell called an 'augmentorium'. The mesh of the screen is calibrated to prevent the flies from escaping, but it does allow their predators to enter. These predators are then scattered over the plot. Three months of this treatment is enough to eradicate vegetable flies. His work has won national awards.[15]

Knowledge of Deguigne's findings reassured these island farmers to transition to bioinspired practices. The results are very positive. They have even created the first organic market in Réunion for certain products.

Conventional Agriculture Weakens Plants

Deguigne's story illustrates the fact that plant protection is not just a matter of technique, it also involves the way we harness nature. The reluctance of conventional farmers to convert to organic farming is partly fuelled by a fear of diseases and pests flourishing. This fear of nature gives rise to a level of control that ends up becoming counterproductive. The more humans intervene, the more agrosystems become vulnerable to bioaggressors and require ever higher doses of pesticides. A death spiral takes hold but farmers do not realise that they themselves are amplifying it. These phytosanitary problems and the cost of chemical treatments affect the environment and the health of consumers, as well as the economy of farms.

Egg laying of a seven-spotted ladybird.

Ladybird larva

One problem that is still deliberately concealed is the effect of pesticides on the health of farmers. I personally lost a friend to a disease related to poorly managed pesticide applications that left him with a serious illness. A mismanaged accident left him in a coma when he was only twenty-one years old, leaving him ill for the rest of his life, before depression drove him to suicide. Classification of diseases linked to the side effects of pesticides as occupational work hazards would result in medical compensation for sick farmers. Their families are also paying a high price. Breast cancer is more common among farmers' wives. A rural doctor told us that he observed a sharp increase in micropenis among farmers' sons. No one stands to gain from the continued use of chemical poisons, except the petrochemical companies.

In the United States, a dramatic increase in the use of pesticides has not been able to stop crop losses.

As Janine M. Benyus points out, "Since 1945, pesticide use has increased by 3300%, but total crop losses due to pests have not decreased. In fact, even though we've bombarded the US with a million tons of pesticides every year, crop losses have increased by 20%. Meanwhile, more than five hundred pests have developed resistance to the most powerful chemicals. Soil is becoming less and less productive. To attempt to remedy this, we have added 20 million tonnes of nitrogen fertiliser per year, or no less than 72 kilos per person in France alone."[16]

Mainstream farming practices contribute to sick plants:

- Monocultures do not exist in nature. They offer bioaggressors gigantic meals of their favourite foods.
- A denatured environment, including the disappearance of trees, hedges, ponds, and plant and animal biodiversity no longer offers shelter or food resources to useful auxiliaries. Populations, harmed by pesticides, are in decline. In France, for example, the loss of birds is dramatic – 30% of bird populations have disappeared from the countryside in the last fifteen years, and 80% of flying insects have been eradicated in the last thirty years.[17]
- Soil, regularly assaulted by machinery, chemical fertilisers, and pesticides, no longer provides healthy food crops. Poorly nourished plants become fragile.
- Plant varieties selected to be hyper-productive are more fragile and dependent on fertilisers and pesticides.
- Chemical fertilisers and pesticides change the metabolism of plants. This point will be developed on p.42.

By disrupting the subtle balance of nature, we have encouraged the development of crop pests.

Is Industrial Biotechnology the Solution?

Disease and pest control is a concern for organic farmers. The trend is towards the industrialisation of cultivation to meet demand. It is common to see large monoculture crops grown on frequently tilled land. Apart from excessive use of chemical molecules, there are not many differences between 'industrial organic' and conventional agriculture.

Changing how we Interact with Nature

The solution lies in natural environments, to take nature as a model instead of being afraid of it. Instead of fighting against countless opponents, let's make the most of the forces at work.

Ecosystems are home to a great diversity of plants and animals. Food chains induce multiple predatory relationships, but a balance can be established or some species will disappear. In the Amazon, I was struck by the genius of the plants and the phytophagous insects that consume 40% of the forest's leaves. Everywhere you look, you can see leaves bit by mouths or mandibles. Plants are constantly inventing new strategies to protect themselves from phytophagous insects. The ingenuity seems limitless. For example, some plants produce toxins that are active only in sunlight; the phytophages, in turn, have become nocturnal and spend the day curled up under leaves. The young shoots of some plants are red so as not to attract the attention of phytophages that consume green leaves. Plants with hollow stems offer shelter to colonies of ants that protect them from undesirables. Plants and their predators have co-evolved over time, a dynamic equilibrium has been established, and contrary to appearances, there are many more cooperative and symbiotic relationships than predatory ones. Alas, the forest remains incredibly lush and full of vitality. Nature's intelligence is awe-inspiring.

Agriculture inexorably leads us to disturb natural environments, if only because we intervene in the soil, select and favour certain plant and animal species, and export produce. Is it any wonder that our interventions lead to imbalances? This is absolutely unavoidable and we must accept it. Instead of dreaming of a pest-free garden, we seek to promote a new balance that allows pests to express themselves positively, without their level of predation endangering our crops.

In nature, diseases and pests are present, but a sense of radiant health prevails. When you walk through an unspoilt environment, you do not usually see sick or dying plants or animals. Predators and diseases affect only the weakest animals and plants, which succumb and disappear. Bio-pests perform an essential function by eliminating weak plants, while resistant plants thrive. In this case pests contribute to the vitality of the ecosystem.

Should we try to eradicate pests? This is impossible and probably not desirable. We could change the way we look at them and see them as allies rather than enemies. If a crop is destroyed by pests, it is not 'their fault', but rather our responsibility. Let us take responsibility for the effect of our actions. We failed to create the optimal conditions for plants to thrive and weakened them. Why did we do this? How did we do it? At what stage? Have we disturbed the soil too much? Is the food our crops extract from it balanced? Rather than bringing out the heavy artillery and cursing these 'pests', let's try to understand how we can improve our practices. When perceived in this way, bioaggressors serve as warnings, as alarm signals, inviting us to make constant progress. This is a complete change of perspective from mainstream agriculture.

A Holistic Approach to Health

There is a profound similarity between conventional agriculture and the allopathic medicine that prevails in the West. Both are symptom-centred. Disease is seen as an anomaly that must be fought. We try to eradicate its manifestations of fever, pain, inflammation and tumour, rather than questioning its causes. The farmer and the doctor try to control the complex systems of the farm and the human body. If a red warning light starts flashing in their control room, they turn it off, without necessarily realising that if the problem that caused the system to go on alert persists, other warning lights will come on. Drug and chemical treatments may turn off the warning, but often their action unbalances other parts of the system. Without disputing the immense contributions of contemporary medicine, it must be said that, despite its impressive progress, the diseases of civilisations are constantly on the rise. Modern agriculture and medicine focus on symptoms and pay little attention to the environment. Moreover, the expertise of the 'experts' too

Ladybird in flight

Cabbage butterfly

Green grasshopper

▲ *Passionate about water, we dug many ponds and renatured the river. This immediately led to an explosion of populations linked to aquatic environments. Many of them are beneficial.*

▸▸ *The photos opposite show:*

Fénoua preparing to release a tench into the newly created pond, rehomed from a farm pond.

White-clawed crayfish are numerous in our ponds, indicating clean water.

*At least 10 species of dragonflies and damselflies are present on the farm, including the rare blue dragonfly, the southern damselfly (*Coenagrion mercuriale*), which has earned our valley its Natura 2000 classification.*

Garter snakes are permanent guests in our ponds and gardens, where they regulate the rodent population.

often leads patients to feel powerless, dispossessed of their health. Similarly, farmers tend to become the executors of directives coming from the chamber of agriculture or the technical salesmen of phytosanitary firms.

Conversely, ecoculture takes the same holistic approach as Eastern and natural medicines. The human body and agroecosystems are highly complex environments with a dynamic equilibrium, self-healing and self-regulating mechanisms. In Chinese medicine, disease or mental pathology is understood to be a manifestation of inner disorder. These signals are precious because they alert us to a deep imbalance and provide information. Doctors look for the causes of the disorder. In the traditional medicines of indigenous Africa, healers involve the patients as well as their relatives, because a pathology is the sign of a disorder in the community, in relationships between people. These forms of holistic medicine are highly effective. They place great emphasis on prevention and the environment. The patient is the main actor in their own recovery. These are practices based on the conviction that good health is our natural state, and our body is endowed with the ability to heal itself.

Plants Heal Themselves

Plants cope with stressors on their own, especially when they are healthy. When a part of a plant is attacked, it can carry out 'cell abandonment', deliberately causing the death of cells in the threatened area. The cells are stimulated by phytohormone messengers released in the second stage of the healing process, with various substances that stop the spread of the disease. This phenomenon of compartmentalisation is easily observed as purplish red spots, for example. In addition, plants can diffuse a phytohormone messenger to all of their parts with various substances that fight the aggressor, blocking oviposition or moult activity. They can also release fungicidal properties. Plants can send messages to neighbours, who can also secrete toxins. These reactions are based on the plant's recognition of the chemical and electrical frequency of the pest, the result of a long co-evolution. If the plant does not recognise the frequency of the pest, it is helpless. This is the case for downy mildew, a fungal disease that appeared in France around 1860.[18]

In light of this recent knowledge, we can ask ourselves if it **is better to grow weak plants and intervene frequently to try to protect them, or strong plants and let** them fend for themselves against their aggressors?

Towards Balance

Ecoculture relies on knowledge, observation and prevention. Crop damage caused by pests is normal and acceptable as long as it does not exceed a certain threshold. Beyond that, pests alert us to the fact that balance has not yet been achieved. This is by nature holistic, and requires attention to many parameters.

The human psyche is such that we tend to divide the world into good guys and bad guys, and position ourselves as victims of the latter. In doing so, we lose some of our power as human beings. Ecoculture suggests another approach. There are no good or bad plants, good or bad insects, but many forms of life that coexist with us in the garden, each performs a function and deserves our respect. We transform space to feed ourselves, and in doing so we have taken responsibility for a little world and its inhabitants. It is up to us to lead it towards dynamic equilibrium. It will never be in a state of absolute stability because countless parameters fluctuate. We must listen carefully to the messages that the garden and its inhabitants send us, and take great care of everything that can contribute to health. Prevent rather than cure. We also have to give up our desire to control everything and trust that every natural space tends towards balance and is endowed with self-regulatory capacities that we barely have any control over. In this way, both garden and gardener will always improve.

The Founding Principles of Garden Health

We have already touched on almost all the factors that promote garden health. However, we will briefly highlight prevention, the basis of our approach, and curative actions that can be taken if necessary.

A Complex Environment

Natural environments have greater complexity, where each ecological niche is fulfilled by animals or plants that have adapted well. Trees, hedges, forest gardens, ponds, crops, and wilderness areas combine forces on the farm to accommodate a diverse number of animals and plants. A dense network of relationships is woven. Each function is carried out by several elements and the ability to self-regulate is strengthened. The whole is more than the sum of its parts.

Complex environments host ecosystem services that play a major role in the health of the garden. Plant nutrition, disease and pest control, pollination, and microclimate creation are some of the benefits. Let us stress once again that 'complex' does not mean 'complicated', quite the contrary – the system tends to self-regulate without intervention.

Towards Stability

In volume three, we described how unstable ecosystems favour animal populations that complete their development cycle quickly. Their evolutionary pattern leads to large population fluctuations. Most crop pests are such species – they are small, have early sexual maturity, little or no parental care, many offspring and a short life span. Rodents, aphids and mites are examples of this.

Conversely, stable ecosystems favour larger, slower-growing and later-maturing animal species with few offspring, sustained parental care and long life-spans. The ecology of the majority of beneficials is based on such a strategy.

By promoting garden stability, we limit the development of pests. If we compare a hectare of motorised organic market gardening with a hectare of eco-growing, of which 1,000 square metres are dedicated to raised bed market gardening, it is easy to see that a ploughing operation will profoundly disrupt the cultivated plot. For an ecoculture plot, the most that needs to be done is decompact a few dozen square metres of permanent beds per day, a gentle operation that has no destabilising effect. Pest management is a recurrent concern in the first case, and a minor issue in the second.

Biodiversity

The number of ecological niches that a garden offers is directly related to the biodiversity that it can accommodate. Biodiversity is a key factor in crop health. Pests are controlled mainly by the presence of numerous useful auxiliaries, including insects, birds, amphibians, snakes, bats, hedgehogs, and birds of prey. This subject is so important that we will devote the next chapter to the hosting of biodiversity.

Soil

Healthy soil produces healthy plants, while unbalanced soil cannot produce perfectly healthy plants, and all the treatments in the world will not help. Would you be healthy if you ate all your meals in a fast-food restaurant? Before tending to crops, the eco-cultivator tends the soil.

The permaculture approach to crop health is aimed at prevention and a dynamic balance between pests and their predators.

Eliot Coleman worked many years before his crops began to show exceptional health with a balanced soil. When he came to Bec Hellouin Farm, Eliot told us that on several occasions he had to remove rocks from some of his beds and, subsequently, pests attacked plants growing only in the area where the soil had been disturbed by this operation. This shows how tilling can damage soil balance and crop health.

By 'healthy soil' we mean soil that is undisturbed by mechanical traffic, rich in organic matter, supple, aerated, and alive, to give plants the nutrients they need. A lack of potash or certain trace elements disturbs the biological functioning of the plant and makes it vulnerable.

A dependable water supply is also a health factor for crops that become weakened by water stress. A soil rich in organic matter retains water and allows roots to grow deep.

The quality of a plant's root system has an enormous influence on its vitality. Many gardeners have observed that vegetables that are attacked have underdeveloped root systems.

Permanent beds provide the best possible environment for crops. The soil, covered with organic matter, becomes healthy. We notice a marked difference between soil quality in our beds and soil in our animal-drawn plots that are regularly disturbed.

Adapted Plants

A stressed plant emits signals that bioaggressors can perceive at great distances, up to several kilometres away. Conversely, the choice of plant varieties that are well adapted to the soil and climate context of the garden is a big factor for success.

▸ *To promote biodiversity in the greenhouse, ponds, flowers, chickens, and perennial vegetation contribute to vibrant and healthier vegetables.*

Hardy and local varieties are generally the most resistant because they have co-evolved for a long time with the bioaggressors present in the region and have adapted to its conditions. The 'formula 1' varieties offered by breeders and exotic varieties often give disappointing results. It takes time to identify the ones that work best for you. Identify them and build up your collection year after year.

Well-cared-for Plants

The care we provide at each stage of a plant's life contributes to its health – sowing in beneficial substrate, at the right time, taking good care of young plants, careful transplanting, watering, weeding, and more. A young plant that suffers from water stress before being transplanted into compacted soil will be more vulnerable to pest attacks than a plant that has grown in the best conditions. Micro-agriculture helps us become attentive to plants.

Banning Chemical Products

Chemicals are strictly forbidden in organic farming. For readers who might be tempted to use them, even as a last resort, beware that synthetic fertilisers and phytosanitary treatments are often accompanied by biological imbalances and an increase in pests like aphids and mites. As Francis Chaboussou's work has shown, pesticides, and in particular herbicides, disrupt the synthesis of proteins within plants, resulting in a dominance of proteolysis over proteosynthesis. Cell walls become thinner and susceptible to free amino acids. This reduces sugars and nitrogen levels in plant tissues.[19] These soluble elements feed insects, fungi, bacteria and viruses, and promote their growth and development. The plants become more palatable and increasingly vulnerable.

According to Chaboussou, **about 80% of plant diseases are caused by inappropriate agricultural practices, especially the use of soluble fertilisers and plant protection products.**[20]

Combining Plants

Plant associations are an effective way of preventing diseases and pests because the potential target plants are mixed together, causing both olfactory and visual interference. Scattering plants such as wormwood, tansy, sage, lavender, mugwort, helichrysum, that give off strong and complex odours, increases this olfactory confusion.

Some plants can also act as physical barriers to the progress of pests. Rows of beans or maize, for example, hinder the movement of carrot flies, which cannot fly very high.

Aromatics can have a positive effect on the health of their immediate neighbours through allelopathic effects (see volume two), although this is still poorly documented.

Some plants act as host plants for pests – the attraction of aphids to broad beans or wormwood is well known. By installing host plants on the periphery of the garden, 'pest cultures' are created that add benefits when the garden needs them. This practice may frighten some gardeners, but most pests are plant-specific, whereas beneficials are usually generalists. Nettles, for example, are specific hosts to the nettle aphid. The hazelnut tree attracts the hazel aphid. The same applies to barley, rowan and elderberry. On the other hand, hoverfly larvae and ladybirds eat aphids from all plant species.

Garden Maintenance

Discard plant parts that become diseased, or the entire plant if necessary. The same applies to fruit trees that can be cleaned up by gentle pruning. Sick or diseased fruit that has fallen to the ground should be collected, unless there are chickens or livestock in the orchard to eat it.

Mulching with Fragrant Plants

We often use aromatic plants as mulch around our gardens. Bioaggressive insects identify targeted plants by sense of smell. They can recognise about thirty different odorous molecules while lavender contains more than 200 different molecules, tansy about 100, and sage 180.[21]

Crop Rotations

Crop rotations help to avoid an increase in pest populations and diseases specific to a particular crop. The large number of crops grown each year in each bed serves as a rotation; however, as far as possible, we avoid having the same plants return to the same bed year after year.

Green Manure Crops

Green manures help clean up soil. Some plants have a specific action against undesirable organisms such as nematodes. Growing green manures from a different botanical family than the crop that follows breaks the development cycles of specialised pests.

Physical and Biological Protection of Crops

Insect Netting

Some plants are very palatable to their pests and the preventive measures mentioned above will not protect them completely. Some market gardeners cover almost all their crops with insect netting in summer. Personally, I don't like to use these unsightly nets. Our carrots and leeks are not protected, yet they are almost free of carrot and leek flies, probably because of crop combinations. We only use anti-insect nets for Chinese cabbage crops in the summer, as they are quickly attacked by flea beetles. You can judge for yourself whether or not it is appropriate to use them depending on the level of infestation.

Traps and Sticky Lures

We use yellow traps to catch flying insects and limit populations, especially in the greenhouse. These traps are made of yellow plastic sheeting, a colour that attracts insects, and covered with a film of glue. They are easy to set with small sticks or bamboo. These traps have the disadvantage that they also capture some beneficials.

Pheromone Traps

These traps attract certain insect species by inducing sexual confusion. We have no experience with these.

Curative Measures

When a pest species or a disease causes damage that is perceived to exceed the tolerable threshold, several options are possible. We can accept the loss of a crop and hope that in time the beneficials will settle in, or that our soil treatments will eventually ameliorate the situation. This approach is quite suitable for farmers who grow small series of different crops. Even if a crop is lost entirely, the impact is limited. It is less acceptable for market gardeners who grow just a few crops on a large scale, as the loss of income becomes significant.

Another route is to try to regulate the problem. There are several possibilities.

Limiting Pest Populations

Our actions can favour certain bioaggressors without their predators being present in sufficient numbers to regulate them. Releasing caterpillars, beetles, slugs, or trapped voles

▲ *Cats are predators of small rodents, such as voles, but their population should be regulated by neutering them, so that they do not reproduce uncontrollably. Cats are also predators of birds, but birds are more difficult to catch than rodents.*

▲ *Hedgehogs are consumers of slugs. Our farm is a hedgehog sanctuary, which means that people who find abandoned young bring them to us for breeding before releasing them into the wild.*

▲ *Despite the proximity of forest, hares and rabbits no longer visit the farm territory, probably due to the presence of our dogs and cats. But in the early years, we sometimes found litters of rabbits even in the greenhouse!*

Slugs and snails are our main concern, but they are no longer a major concern, just a point of vigilance, due to the presence of many useful helpers.

will help control the population of pests. The relevance of these interventions must be assessed on a case-by-case basis – if the crop is nearing its end, it is not necessary to intervene. Colorado beetles are only a threat if they eat more than 25% of potato foliage. At Bec Hellouin Farm we only intervene for slugs and voles – we will deal specifically with these two pests at the end of this chapter.

Misting with Black Soap

Some pests such as aphids and mites can be disturbed by short and repeated spraying of foliage with water. Adding black soap to the water is more effective.

Perrine's Preparation

Pepper plants are highly attractive to aphids. Before transplanting, soak them in the following preparation – 1 litre of rainwater or river water, 5 tablespoons of pure black soap and 1 tablespoon of bicarbonate. Soak the entire plant in a bucket of this solution for a few seconds. The black soap removes aphids and makes it difficult for them to spread, while the bicarbonate prevents diseases from creeping in where the aphids have weakened the plant.

Introductory Biological Control

This technique consists of purchasing and releasing large quantities of a given beneficiary in a plot, preferably a greenhouse, to contain a particular pest. Some beneficials, like parasitoid wasps, are capable of regulating a pest outbreak. Bacteria such as *Bacillus thuringiensis* penetrate insect pests to destroy them from within.

Treatments based on the introduction of entomopathogenic fungi or nematodes that attack insects can also be used.

We do not take this approach on our farm. We prefer to create areas of biodiversity right in the heart of the greenhouse. In early spring, Perrine collects ladybirds from the nettles in the forest garden and introduces them where they are needed during the first aphid attack of the season.

Bordeaux Mixture

Bordeaux mixture is copper sulphate plus lime, a copper-based preparation approved for use in organic farming against mildew. It is generally accepted that copper has an antifungal effect, but it may be that this trace element strengthens plants and helps them overcome disease. We use it against downy mildew in tomatoes. Some growers are reluctant to use it for fear that a build-up of copper will harm the fungus populations of soil. To avoid this risk, it is best to keep applications of Bordeaux mixture to a minimum. A late-season attack of mildew on tomatoes is normal and may not require intervention, whereas a spring or summer attack should be dealt with very quickly. However, as part of a research programme on compost, soil tests were carried out month by month in our greenhouse. They showed that there was less copper in our soils at the end of the season than in the spring, when two or three sprays of Bordeaux mixture had been applied. This probably means that the crops had consumed more copper than we had applied with the Bordeaux mixture.

Fermented Extracts and Plant Decoctions

Fermented extracts, infusions, macerations, decoctions of certain plants such as nettle slurry, comfrey, wormwood, horsetail, rhubarb, garlic... are appreciated by organic gardeners. Only the fermented fern extract seems to have a repellent effect on certain pests like aphids. Other extracts are mainly intended to stimulate the growth to boost a plant's defences. As these nutrient concentrates can also feed pests, it is best to use them as a preventive measure.

There are many preparations available. Éric Petiot's excellent book *Purin d'ortie et Compagnie* is a great reference on the subject. There are also many resources on the internet.[22]

Essential Oils, Sugar, and Honey

Some growers have been experimenting with curative treatments based on essential oils. We have tested some of these with good results. Extremely low doses of honey or sugar boost the microbial life of soil.

Effective Microorganisms, Compost Tea

When a crop is attacked, Perrine sprays effective microorganisms, or compost tea. These preparations do not seem to have a curative effect, but they can help strengthen plants and are beneficial for soil life.

We find that we do not have enough time each year to make and spray these herbal or microorganism preparations.

To make this easier, we have installed two tanks at the entrance to the greenhouse for their production. They are located on a platform and equipped with a valve, which allows buckets to be filled directly. A dosing pump, activated by water, allows the preparations to be distributed via the greenhouse's irrigation system at a dosage of 2%.

Mammals and Birds

In a diversified agroecosystem like ours, mammals such as voles, coypu, rabbits, possibly deer, wild boar, and pigeons are more of a problem. Good fences keep us safe from the inhabitants of the forest that borders the farm. Our two dogs and four cats naturally control rabbits and voles. However, we are careful not to let the cats proliferate as they also attack birds - we neuter after their first litter. We trap voles in the greenhouse and coypu, which enjoy our island garden. Pigeons cause a lot of damage to young cabbages and it is difficult to keep them at bay. We use scarecrows, such as those mentioned in Chapter 9, p.93. Sometimes a buzzer that imitates the call of birds of prey is used. When pigeons take up residence in the tall trees around the gardens, I sometimes take out my grandfather's rifle and fire a few rounds into the air. They quickly realise that the area is potentially dangerous.

Slugs

When we adopted the permanent bed technique in 2009, slugs were a big problem for four or five years, and we remain vigilant. Mulched beds favour them, as well as voles. We use a strategy based on several elements:

- No **mulching** of beds in the spring, when slugs are most active and young plants are most palatable.
- No mulches for vulnerable crops such as lettuce and Chinese cabbage.
- Use abrasive mulches that slugs do not like, such as bark.
- **Wood ash** - although it is difficult to apply for a large garden, not very effective and may unbalance soil if applied repeatedly.
- Trapping - with **beer traps** (not very effective in a large garden) or boards laid on the ground, which are turned over in the morning to collect the slugs that have taken shelter there.
- **Collecting** slugs in the morning in wet weather is probably the simplest and most effective course of action.
- Occasional introduction of **Indian runner ducks.** These are an effective way of keeping the slug population down, but as adults they can also feed on young plants and cause significant damage. It is best to keep them in the garden for only a few hours a day, which requires a strong dose of vigilance.
- Good use of **hens:** our poultry have runs around the market gardens and they feed on slug eggs and larvae. As these grassy areas are potentially reservoirs of gastropods, the hens effectively sanitise them.
- Encourage **natural predators:** batrachians, carabids, hedgehogs, etc. They do most of the regulating once an ecological balance is reached in the garden, which can take several years.
- As a last resort, we apply **a slug control product** that is authorised for organic farming called Ferramol, or Sluxx, which are similar in composition - iron phosphate - but Sluxx is more concentrated (photo on previous page). These products are applied on palatable plants when slugs are active. We mainly use it in the greenhouse, in spring, to prevent the first slugs of the year from breeding and then maintain it for the season as needed.

The combined effect of these different actions is to keep predation at an acceptable level.

Compliance with Legislation

Legislation governs the control of diseases and pests, including the use of 'low concern preparations' such as plant extracts. The natural preparations described in this chapter are not necessarily all authorised in France today. Market gardeners must keep a phytosanitary certificate of their interventions.

Mainstream agriculture simplifies agrosystems and processes. For each problem, it relies on one or more powerful set of 'miracle molecules' that eradicate the problem, at least in appearance. In natural agriculture, we do not have a 'miracle molecule' or a powerful product. We use a whole range of solutions, gentle practices and small actions that work in synergy. The sum of these small actions leads to greater efficiency. This approach is based on knowledge of nature and respect for the environment and its inhabitants. Permaculture practices allow the garden to gradually achieve the dynamic balance that we actively seek.

▲ *Ground beetles, toads and frogs, hedgehogs, chickens and Indian runner ducks are our main allies in controlling gastropods.*

◂ Year after year, there is more wildlife on the farm. It gives us an enormous sense of satisfaction to produce abundant amounts of food for humans while increasing biodiversity. However, it took years for owls and other birds of prey to become regular, if not permanent, residents of the farm.

5

HOSTING BIODIVERSITY

Permaculture aims to develop the resilience of individuals, regions, societies and ecosystems.

Gildas Véret[23]

Attracting significant amounts of wild animal and plant species is critical for balance and resilience. It is also a commitment to the biosphere, which is currently facing its sixth mass biodiversity extinction, the last being dinosaurs 65 million years ago. Scientists estimate that in modern times, 7% of species have probably already disappeared, and that this figure could rise to 25% if nothing is done.[24]

If widespread, ecoculture could play a major role in preserving biodiversity for two main reasons. The first is that by diversifying and increasing the complexity of our micro-farms, they provide shelter for many species. The second is producing more food in less space, which could potentially free large areas of land from human control to become biodiversity reserves.

Agronomists and naturalists have been studying the impact of our agricultural practices on the farm's biodiversity. The initial results have been very encouraging. There are more earthworms, birds, wild bees, and aquatic animals on our farm than in surrounding cultivated areas. The pond surrounding our island gardens is home to white-clawed crayfish, a species whose presence attests to the quality of the water, and no less than ten species of dragonfly, including the lovely blue damselfly.

Accumulating biodiversity has become a major focus at Bec Hellouin Farm. We have been officially recognised as a bird sanctuary and we partner with a hedgehog refuge. People who find a sick hedgehog or an abandoned litter can bring them to us. I have often admired Perrine's generosity, getting up at night to feed little hedgehogs, while her days are so full.

Permaculture Design

The design of the farm encourages biodiversity. When we first arrived there was just a bare meadow and a river; now the farm has a mosaic of different habitats. Comparative studies have shown that wild bees are found in greater numbers in our less well-maintained gardens, which offer them more resources than the zone 1 gardens. The agronomists who carried out these studies advised us to take the following measures to boost biodiversity:

- The concept of zones in permaculture design is often reduced to organising spaces in concentric circles, but it is better to have interlocking zones – juxtaposing the highly manicured gardens of zone 1 with the wilder areas of zones 4 and 5; even if the latter are small, they will be more beneficial, especially when linked to larger wild areas.
- Introduction of trees, ponds, flowers, perennial plants, nest boxes into the heart of the most anthropised spaces, such as the greenhouse.
- Link to biodiversity corridors, connecting the intensive heart of the farm to nearby forest for the free circulation of auxiliaries.

We will now look at how biodiversity contributes to the health of the garden, and then consider what can be done to support it.

I wonder what our farm will look like in ten or twenty years, how it will react to climate change, what wildlife will find refuge there.

Beneficial Insects

The term 'beneficials' covers several groups of animals that are helpful in the garden:

- **Insects:** ladybirds, lacewings, ground beetles and staphylines, hoverflies, spiders, predatory mites, predatory bugs, to give just a few examples.
- **Amphibians:** frogs, toads and much more.
- **Reptiles:** slow worms and snakes, for example.
- **Mammals:** hedgehogs, cats and dogs – these are predators of rodents and rabbits.
- **Birds:** insectivores, especially passerines and birds of prey.

Beneficials can be grouped into several categories:

- **Natural enemies**, predators and parasitoids of pests, which play an essential role in crop protection. These natural enemies can be:
  - **Generalists:** they protect crops but are unable to manage an outbreak like that of ground beetles.
  - **Specialists:** they clean up and are able to manage an outbreak, parasitic wasps, for example. A single female hoverfly can eliminate 300,000 to 1 million aphids.[25]
- **Pollinators**, essential for fruit production in the botanical sense.
- **Decomposers of organic matter** contribute to the fertility of the agroecosystem and the carbon cycle.

There are redundancies between certain functional groups. Hoverflies, for example, are both natural enemies when the larvae eat aphids and pollinators when the adults consume pollen. It is the larval stage that is of interest for crop protection from hoverflies and other pests.

Natural enemies often have a mixed diet:

- Larva is carnivorous and consumes phytophages.
- The adult consumes nectar, a source of sugars, pollen, a source of proteins, and honeydew, a source of sugars and an indicator of aphid presence that makes favourable conditions for egg laying because larvae will easily find food resources there.

The notion of auxiliary should be put into perspective. For example, the earwig is an aphid predator. It is, from this point of view, an auxiliary. However, if aphids become scarce, earwigs can attack the fruit… and then become a pest! The easiest way to rectify this is to trap them in small upside-down pots filled with hay, and then move these pots to areas of the garden where aphids are present.

Moles are excellent soil workers. They perform valuable decompaction and aeration tunnels, sometimes at the expense of seedlings and young plants that can uproot in their trail. Even slugs have interesting functions. They decompose organic matter, an essential starting point for carbon recycling.

It is helpful to understand and know the role of each beneficial in the ecosystem, including their lifestyle and habits.

▼ *The various species of chickadee, including the blue tit pictured here, are permanent residents of our gardens. In winter we have feeders outside our windows. I suppose our little friends would do fine without them, but it gives us the joy of observing them up close.*

Encouraging Beneficial Organisms throughout the Year

In order to encourage allies, an ecological approach to their habitat, their feeding and their reproduction must be put in place. The objective is for beneficials to find the conditions necessary for their survival in the garden throughout the year. Blaise Leclerc and Gilles Leblais identify **six basic rules:**[26]

1. Feed the beneficials properly before winter arrives (late bloomers, seeds, last prey).
2. Provide overwintering shelters.
3. Feed them at the end of winter (early flowering, first prey).
4. Provide them with flowers all year round without interruption.
5. Provide relay plants for their favourite prey.
6. Provide water during dry summers and winter frosts.

Feeding Beneficials

Plant diversity ensures that beneficials have food available all year round and vegetation also gives them shelter.

Species diversity is most commonly found in areas with native plants, trees and shrubs that are well adapted to soil and climate. Native species have long co-evolved with animals and provide them with the shelter and food they need. In contrast, many horticultural varieties, selected for their beauty, no longer produce pollen, nectar or seeds. Exotic varieties can be invasive. Plants are chosen for uninterrupted year-round bloom and 24-hour feeding as some beneficials have a nocturnal life and need access to flowers that open at night, like the tilted silene, *Silene nutans*.

Most beneficials like simple, shallow flowers. Flowers of the Apiaceae family including umbellifers such as green anise, dill, hogweed, carrot, chervil, celery, parsley, parsnip, fennel, coriander, lovage, caraway, angelica, rockrose, hemlock, and garlic are particularly attractive. Their umbels are attractive to hoverflies in particular. Bumblebees and butterflies can forage on deeper flowers.

The presence of beneficial insects can also be supported by sources such as vegetables that have gone to flower and seed, rotting fruit, decomposing organic matter like mulch, berries and seeds for birds. During cold spells, birds appreciate feeders. If you place them in the middle of a window, they are a fascinating sight. However, be sure to keep them out of reach of cats.

Shelters that Encourage Beneficial Organisms

A diversified plant environment with wild sections and a wide variety of ecological niches offers the best possible place for beneficials. They are particularly vulnerable in winter and appreciate simple, natural shelters. These shelters can be forests, wild trees, hedges, bushes and copses, thickets, fallow land, meadows, perennial beds, ground cover plants, mosses, climbing plants, green manures, and more. Old hollow trees, stumps and dead trees are valuable and should be preserved.

Evergreen plants like ivy and holly, and marcescent plants that keep their leaves in winter, such as young hornbeams and beeches, are a valuable resource.

Market gardeners will benefit from offering non-plant habitat to allies including piles of stones, flat stones, drystone walls, various cavities in the ground and in buildings, boards and tiles laid on the ground, and mud. Ponds, pools and troughs are essential, especially in hot summer weather and during winter frosts.

Nesting boxes and insect shelters or hotels can also be installed. There are a wide variety of different shelters available for birds, ladybirds and other insects, hedgehogs, bats and even weasels. These artificial shelters are not necessarily ideal and require annual maintenance. However, we have installed quite a number of them on the farm. Extensive documentation and plans are available on the internet.

The thatched roof of our cottage is an excellent shelter for insects and even small rodents, to the dismay of our youngest daughters!

How Beneficials Circulate

The circulation of beneficials from the wildest to the most intensively cultivated areas needs biodiversity corridors that link them. These can be a variety of hedges, with a ground clearance of at least 3 metres, planted along with two or three rows of trees. Grassed alleys are also a route for beneficials.

In the absence of a biodiversity corridor, the distance between wild areas and gardens should be close enough to allow the movement of beneficials. A distance of about 50 metres seems to be a maximum.

Perrine and I have really noticed a change in biodiversity since we started transforming the farm. Our wild companions are a source of wonder every day.

One of the most exciting aspects of our farming adventure is to see it densely occupied by humans and wildlife. During training weeks there are more than 60 people in a small area, and during business days more than 600 can be present, and yet the farm remains a welcoming environment for wildlife.

▲ *This old hollow ash tree sits on the riverbank, just across from my office. Each year its cavities serve as nesting boxes for various species of birds, such as the lovely nuthatch whose acrobatics often entertain me. These old trees are excellent shelter for biodiversity, and they absolutely must be protected.*

▼ *Following the advice of our naturalist friends, we have installed a large number of nesting boxes of various types on the farm. This one is a bat nesting box. Bats eat a lot of insects.*

Goldfinch

Spotted woodpecker

Grey wagtail

Birds, precious allies

by Camille Joyeux, agricultural engineer, doctor of pharmacy[27]

Our most precious feathered allies are passerines, or songbirds. Tits, nuthatches, treecreepers, redstarts, grey flycatchers, black flycatchers, starlings and more brighten up our gardens with their songs and colours. They are the most widespread group of birds, characterised by their four toes that oppose another three, and by their pointed beaks. Fed from the nest, the chicks claim their food by opening their beaks.

Birds that frequent gardens have a variety of feeding habits. In autumn and winter, they eat wild berries and fruits, like yew, honeysuckle, privet, black elder, blackberry, dogwood, viburnum, and sea buckthorn. Their nut consumption includes hazelnuts, acorns, seeds of Lamiaceae, conifers, maple or birch. During the summer, their diet is based on insects, spiders and small molluscs. Chicks are raised during the months of the year that are richest in insects, caterpillars, larvae and eggs, all important sources of protein for their development. All garden birds are insectivorous during the breeding season.

The majority of garden birds are cavernicolous species that nest in cavities such as hollow trees, walls, rocks, for example. This is the case for the great tit, the blue tit, the nuthatch, the black red-tail or the creeper. Some species such as the robin are semi-cavernicoles because they nest in open cavities.

▼ *Birdhouse*

Cavernicolous species are among those most affected by human activities over the last half-century. The abandonment of orchards, the systematic cutting of unprofitable deciduous trees, the demolition or renovation of old buildings, the removal of hedges and brambles, the cleaning of parks and gardens and the standardisation of woods and forests are reasons for the rarefaction of natural cavities for nesting. Cavernicolous birds will settle in the first hole that comes along, whether it is suitable or not, making reproduction uncertain.

In addition to the scarcity of natural cavities, the reduction of food resources for birds is a direct consequence of the destruction of natural environments, the trend towards mono-culture and the widespread use of pesticides.

The manifestations of climate change and the appearance of invasive species further darken this already gloomy picture.

Due to lack of shelter and cover, these precious auxiliaries are now threatened. According to the League for the Protection of Birds (LPO), 13% of the 533 species present in Europe are threatened. In thirty years, more than 420 million birds will disappear from Europe. This news is alarming.

Common snipe

Redwing

Cirl bunting

Nesting boxes

A nesting box is characterised by several parameters, designed for the species of birds that will most likely occupy it:

- The diameter of the hole.
- The shape.
- The size of the nesting chamber and the depth of the nesting box.
- The amount of light in the nesting chamber.
- The height of the nesting box.
- The environment where the nesting box is installed.
- The period of installation of the nesting box.

There are several types. The so-called 'generalist' nesting boxes accommodate different species of birds. Conversely, specialised nesting boxes are intended for a particular species of bird, like the 'nest shelf' for robins.

Nesting boxes are available commercially, and you can make them yourself.

▼ *Nesting box*

Kingfisher

Eurasian bullfinch

Eurasian blue tit

The birds at Bec Hellouin Farm

by Gauthier Chapelle, naturalist, biologist and agronomist

Birds are pampered by the cultivation practices at Bec Hellouin Farm. The conclusions to follow are based both on 'opportunistic' observations made by various visiting naturalists, but also on a spring census covering four days in March and four days in May 2018. This preliminary survey confirmed the regular presence of more than 40 species feeding on the farm, including about 30 nesting species.

While the Bec Hellouin Valley initially had a nice avifauna, the initial pastures would not have attracted many. However, it is very clear that the cultivation practices specific to the farm have multiplied the (micro-)biotopes favourable to a certain number of species. The massive addition of trees and hedges (still being developed), whether in the forest garden, the vegetable garden or around the plots, the extension of ponds and pools (favourable to the red and green frogs, the common toad, the newt and the grass snake, all of which are present), and even the various buildings provide so many opportunities for feeding or breeding (even the greenhouses are used as hunting grounds).

We benefit from birds that are plentiful on the farm, such as moorhen, grey wagtail, swallow, blackcap, blackbird, black redstart, blue tit, Eurasian wren, dunnock and chaffinch. Remarkably, several species that are declining in the countryside due to industrial agriculture are present as well on the farm, such as grey partridge, goldfinch and linnet. The combination of hedgerow habitat, a clean river and the absence of insecticides probably explains the presence and nesting of the more surprising species found here such as mistle thrush, garden warbler, spotted flycatcher and common redstart.

Great tit

Nuthatch

Common redstart

Birds species at Bec Hellouin Farm

Here is a non-exhaustive list of the birds that visit the farm:

Dunnock
Little egret
Common snipe
Grey wagtail
Pied wagtail
Honey buzzard
Bullfinch
Cirl bunting
Common buzzard
Mallard
Goldfinch
Jackdaw
Tawny owl
Black crow
Common cuckoo
European starling
Ring-necked pheasant
Blackcap
Garden warbler
Jay

Spotted flycatcher
Herring gull
Great cormorant
Short-toed treecreeper
Mistle thrush
Redwing
Song thrush
Grey heron
Cattle egret
Barn swallow
House martin
Melodious warbler
Linnet
Kingfisher
Common swift
Blackbird
Long-tailed tit
Blue tit
Great tit
Marsh tit

House sparrow
Grey partridge
Spotted woodpecker
Green woodpecker
Magpie
Wood pigeon
Eurasian chaffinch
Meadow pipit
Common chiffchaff
Common moorhen
Goldcrest
Common firecrest
Robin
Common redstart
Black redstart
European serin
Eurasian nuthatch
Eurasian wren
European greenfinch

Here are some of the wild bees seen at Bec Hellouin Farm.

Anthophora plumipes

Bombus lapidarius

Bombus hypnorum

Bombus pratorum

Osmia cornuta

Andrena bicolor

Andrena haemorrhoa

Osmia caerulescens

Lasioglossum calceatum

Osmia bicornis

The biodiversity of microfarms

by Nicolas Vereecken, engineer in agronomic sciences, doctor in biological sciences and professor of agroecology at the Université Libre de Bruxelles

Intensive agriculture and the decline of biodiversity

Everywhere in the world where intensive agriculture has simplified landscapes, removed woody hedges, and reduced the number of different crops per hectare, the result is always the same: the diversity of wild plants, birds, insects, and amphibians has decreased, sometimes very dramatically. A large part of our natural capital has been wiped out. We are only just beginning to measure the negative consequences on yields and crop stability from one year to the next, and on the quality of our fruit and vegetables.

When we become fully aware of the condition of land damaged by intensive agriculture – in Europe, Africa, America and Asia, and around the world – we can determine if it is not too late to regenerate it, and if it is still possible to reverse the trends on a wider scale.

Networks of biodiversity oases

Scientific research in recent years has shown that the 'quality' of a landscape is crucial for hosting biodiversity. If you set up an oasis in the middle of a desert, biodiversity may take hold, timidly, little by little, but with less vigour than if it were connected to a network of other oases that host biodiversity.

The logic is the same with agricultural land. The impact of biodiversity-friendly developments increases when they are interconnected and take place where natural or semi-natural elements still exist, such as meadows, fallow land, nature reserves, and forests.

Local solutions to global disorder?

If biodiversity is threatened on a large scale, should the solution necessarily be large scale? In other words, can we have a real impact by acting locally, if everyone 'does their bit'? At first glance, it's a bit like David against Goliath... but, on closer inspection, it seems that some very inspiring small-scale regenerative practices are already successful all over the world.

Diversified market gardening on small surface areas has always been a blind spot in agronomic research, and few researchers have taken an interest in the potential of these special types of microfarms for hosting biodiversity. Market garden microfarms produce leafy vegetables, root vegetables, and are partially occupied by greenhouses... microfarms may not seem particularly favourable places for biodiversity. Scientific and agronomic research is not yet well 'connected' to these issues.

Towards a network of biodiverse microfarms

Studies that my team and colleagues have been conducting for several years at Bec Hellouin Farm, and in a network of urban and rural market gardens in Belgium and Canada, have revealed a hitherto unknown function of microfarms – that they make a unique contribution to hosting biodiversity, much more than large farms and as diverse as nature reserves located in the same environment. The results suggest a universal phenomenon that should invite us to reassess the role of microfarms in biodiversity conservation.

Bombus terrestris

Bombus hortorum

Bombus vestalis

Bombus pascuorum

Megachile willughbiella

Andrena fulva

Lasioglossum morio

Andrena flavipes

Andrena cineraria

Chelostoma florisomne

Andrena dorsata

Wild bees do not make honey and are solitary, like butterflies and beetles. They are very diverse, with nearly a thousand different species in France, over 2,000 in Europe and over 20,000 worldwide. Although less known to the general public, farmers and politicians, they provide invaluable services to our environment and agriculture. The microfarms studied to date, in particular Bec Hellouin Farm, host several dozen different species of these essential pollinators throughout the year. This ensures the reproduction of wild and cultivated plants within and outside the farm, contributing to its status as a welcoming 'micro-ecosystem' for bumblebees, *Andrena*, *Osmia* and many other wild bees.

Wild bees in the garden beds at Bec Hellouin Farm

Students Margaux Villebrun, Luc Devaux and Paul Emptaz inventoried the farm's wild bee populations. The results clearly showed a level of complementarity between island garden, mandala garden, and apple orchards located alongside greenhouses. Each zone constitutes a 'small world' with its own unique set of species. All of the beds have about ten species in common that are found in each zone of the farm.

There are more than 30 different species of wild bees on site, including no less than eight species of bumblebee that strengthen links between wild and locally cultivated flowering plants.

Andrena, *Osmia* and spring bumblebees are numerous on the farm. They contribute to fruit tree pollination, and spring plants in particular.

How can biodiversity be promoted on microfarms?

What steps should be taken to foster harmony between microfarming and nature?

Here are a few tips:

- Give preference to perennial plants and strive for variety and a year-round abundance of flowers.
- Opt for 'companion' plants from your region – exotic species are not necessarily in tune with the needs of local pollinators, and some exotic plants might be invasive.
- Reinforce local floral resources by installing beds of aromatic and medicinal plants, in particular Lamiaceae (sage, thistle, lamia, lavender), Fabaceae (alfalfa, vetch, trefoil, clover) and Asteraceae (camomile, tansy, thistle). This list is far from exhaustive!
- Provide shelter for *Osmia* and *Megachile* by installing a wild bee hotel. They can be made with recycled materials.
- Fruit trees and berries provide abundant food for pollinators at the beginning of the season and encourage them to settle on your microfarm. Abundance and diversity of pollinator species are the two keys to successful production and more stable yields from one year to the next.

To recap, the ability of microfarms to host functional biodiversity opens important perspectives on pollination services that are enhanced from this type of production. Biointensive agriculture and biodiversity conservation are interdependent. The FAO has been indicating for years that biodiversity must be at the core of future agricultural practices and all efforts to strengthen food security.

XIII

ANIMALS AND ANIMAL TRACTION

◄ *It is difficult to imagine a farm without animals. Our farm's business is primarily market gardening and arboriculture, but we have always welcomed animals, like this superb turkey. Although work overload has led us to give up goats, guinea fowl, geese and rabbits, we still dream of welcoming a small cow.*

6

ANIMALS

Pigs dozed off with their snouts in the mud; calves bellowed; sheep bleated; cows, with one hock folded spread their bellies on the grass and, slowly ruminating, blinked their heavy eyelids at flies that buzzed around... In the distance [...] a great muzzled black bull with an iron ring in its nostril that moved no more than a bronzed beast. A child in rags held it by a rope.

Gustave Flaubert[28]

There are many advantages to having farm animals. Throughout history, farmers have combined crops and livestock, creating the same beneficial positive interactions between plants and animals found in nature. Animals benefit from plant food and plants benefit from the fertility provided by animal excrement. This makes a farm's ecosystem more diverse and increases energy efficiency.

Conversely, separating crops and livestock with specialised farms and regions is counterproductive. This mindset comes from planners trained at the top schools but lacking any farming experience. The result is an artificialisation of agriculture that harms sustainability.

The Beauce region of France once had fertile plains. They were dedicated to grain crops during the industrial agriculture boom and just a few decades later suffered from animal manure deprivation. At the same time Brittany, a region with an oceanic climate surrounded by ports for animal feed imports, was devoted exclusively to livestock farming. This magnificent region has long been suffocated by slurry and other livestock effluents that pollute its fresh water and coastline.

Mixed farming is an elegant and efficient style of agriculture that allows cycles to be completed on a smaller, microfarm-size scale. Ecoculture takes this a step further by increasing beneficial interactions between trees, crops and animals. However, raising animals brings up a number of societal, ecological and ethical issues.

▼ *This old veterinary guide presents images of a vanished rural life, yet, paradox of our times, it is these kinds of images that we still find in children's books today – they never show battery farms because they are a shame for humanity! Our imagination has been shaped by these children's books; is that why we take so much pleasure in seeing our animals frolic freely under apple trees?*

The Impact of Livestock Farming on the Planet

We humans have a problem: our numbers are increasing and as our standard of living rises, we consume more animal protein. The result is a perpetual need for large-scale livestock farms that shift us away from farming and into industrial processes. To feed these intensive livestock farms, forests are being cleared and land is being desertified. Crops for animal feed are energy-intensive and consume a lot of fresh water – it takes 10,000 litres of fresh water to produce 1 kilo of beef. The impact of livestock on the climate is harmful, especially methane emitting cow burps that add to greenhouse gas.[29] Furthermore, eating too much meat and dairy products is bad for human health.

I witnessed other perverse effects of livestock farming during our round-the-world trip aboard the *Fleur de Lampaul*. In Kerala, a beautiful province in southwest India, we investigated the disappearance of small-scale fishing. Until recently, Kerala was home to several hundred thousand small-scale fishermen who caught fish that are abundant in the shallow coastal waters. They made a modest living in extremely rudimentary boats.

A pirogue boat supports seven times more fishermen per tonne of fish landed than a trawler. These fish, sold at low prices, were the main source of protein for the people of southern India.

Then 'progress came' in the form of foreign capital that made it possible to build several thousand small motor trawlers, along with the

Taking responsibility for raising animals for food raises ethical questions, not least because meat consumption adds significantly to our ecological footprint.

cold storage and industrial logistics needed to process large volumes of catch. Fish caught by trawlers are more expensive than those landed by pirogues and have become largely inaccessible to local populations, who have lost their main source of protein. Fish caught in Kerala is processed into animal meal and exported to feed livestock in rich countries.

Another consequence, an ecological one, is that trawlers have emptied coastal waters of their fish stocks. Small-scale fishermen have been forced to go further offshore, with all the risks that this entails, to bring in meagre catches, or to give up their livelihoods altogether.

This example illustrates the evils of a globalised system that is geared only to short-term satisfaction of demand from those with economic power. The result is the impoverishment of the most vulnerable and environmental destruction, without making the inhabitants of the overfed countries any better off, as they are unable to stop the rise of so-called 'civilisation' diseases, such as obesity and cancers that strike a growing number of children every year. Each of us is responsible for this state of affairs, since we consume meat without knowing exactly under what conditions it was produced. Our overstuffed fridges and overweight bodies almost certainly contribute to the increase in malnutrition in Kerala or the deforestation of the Amazon.

I might add that after leaving Kerala, we stopped off in the Maldives archipelago of atolls that are beginning to be submerged by rising seas, even though this place has only a few thousand cars and no industry. All Maldivians, from small fishermen to the President of the Republic, made a strong appeal for us to unite in the fight against climate change.

Everything is linked; we can choose to be part of the problem or part of the solution. The permaculture approach leads us to break out of destructive spirals and enter into a logic of reconstruction. It is up to each of us to relocalise everything possible and to think about the sustainability and resilience of each system at its smallest level - the one we can act on - by seeking to achieve high energy efficiency through processes that are as simplified as possible.

Is it Ethical to Raise Animals?

Animal husbandry is also a matter of ethics. The choice to kill animals for food is a personal one and we do not claim to have a legitimate answer to this complex question. But let us share with you the position that we, as farmers, have come to adopt.

From time immemorial, almost all humans have fed on meat, first by hunting it and then by domesticating animals. Animals have not always been treated well, as ancient societies did not necessarily acknowledge animal suffering.

However, factory farms, the transport of animals from one country to another and slaughterhouses have taken the barbarity up a notch. All too often, they are a disgrace to humanity. Respect for life has almost always been sacrificed on the altar of profit. In order to ingest meat in large quantities at low cost, we turn a blind eye to a system of unbearable cruelty. We are cowardly accomplices of the executioners to whom we delegate the dirty work by closing our eyes, and we can be sure that ingesting this anonymous meat soaked in suffering does no good to our bodies.

More and more citizens of the modern world are making the choice to become vegetarian or even vegan. We now know that we can improve our health and vitality by refraining from eating meat, while significantly

▸ This Bayeux Normande sow, named Anne-Christine after one of our favourite interns, introduced us to the amazing intelligence and sensitivity of pigs. She would roll over with happiness when Perrine gave her a massage. We could never bring ourselves to put her down. Domesticated 10,000 years ago, pigs have many genetic and physiological similarities to us.

◄ *As an educational farm, we have many opportunities to witness how immediately complicity is established between most children and animals – here is our donkey Alice and her foal. A relationship with animals is a form of therapy.*

reducing our ecological footprint. Studies of multi-generational vegetarian populations in India and the United States have highlighted the benefits of this diet. However, most people who follow this path do so mainly for ethical reasons, refusing to contribute to animal suffering.

The Place of Animals on a Farm

We have the utmost respect for personal choices such as these. We ourselves were strictly vegetarian when we trained in yoga and one of our daughters, Shanti, decided to be vegetarian from an early age, despite all the constraints that this may have posed, being the only vegetarian student in her school. This is how much we admire and respect vegetarianism.

Becoming farmers has made us aware of other facets of reality which have led us, for the time being, to distance ourselves from a strictly vegetarian diet. We allow ourselves to eat animals that are raised with dignity on our farm, slaughtered with respect and gratitude, without physical or moral suffering.

One of the things that has changed our position is our awareness of the many services that farm animals provide on the farm:

- They contribute to cycle completion, by adding value to waste and by-products from the farm.
- They transform vegetable waste and grass into energy.
- They help to maintain areas that would otherwise be wasteland, unless we maintain them ourselves, which would take time and energy. Our meadows, for example, would be lost without them.
- Their droppings improve soil fertility.
- They sanitise orchards by consuming fruit parasites.

They are also a friendly presence on the farm. The singing of the chickens laying eggs in the morning, the tweet-tweets of the chicks running between our feet to grab breadcrumbs, the playful jumps of the lambs, the bleating of the ewes calling them, the friendly greetings of our donkey. Animals give our farm a soul. They teach us a lot, starting with respect.

Livestock as Bearers of Culture

Farm animals are faithful companions who have co-evolved with us for thousands of years. What would our farms and countryside be without sheep, horses and cows in the meadows, without barnyard animals pecking freely around the barns? Farm animals have shaped many of our traditional landscapes and have had a profound effect on art, literature, spirituality and our imagination.

If we were to give up breeding, we would have to wipe out all the local breeds that are so diverse and well adapted to the land that they would be unable to return to the wild. We also have a responsibility towards these animals that generations of breeders have evolved.

Gastronomy is an integral part of the culture of a civilisation and its art of living. In France, we have so many recipes to accompany the products of farm animals that

▲ *We have so many animals on our farm because we enjoy their company. It was a fox that got the better of our beautiful turkey...*

I cannot imagine having to wipe out this heritage.

To tell the truth, there is also self-interest in our choice to raise animals because their meat is so tasty and we enjoy a good meal from time to time, usually when we have friends over. It is also beneficial in terms of health.

A Choice of Conscience

We have witnessed scenes that speak volumes about the sensitivity of our companions, for example when our sow, despite her respectable size, stood on her hind legs and wrapped her front legs around the neck of Alice, our donkey, for a long hug. They were both equally enjoying it. Those who would accuse us of anthropomorphism have no idea how close animals are to us, and we to them. We have a genetic make-up very similar to that of pigs.

In the end, it is up to each individual to take a stand on animal issues. Some farmers decide not to raise any and refuse to fertilise their land with animal products. There is no need to comment on these conscientious choices. The main thing is that everyone finds their own way.

▼ *Perrine and Trottinette, a mini horse who has lived on the farm for fourteen years.*

A Permaculture Approach to Animal Husbandry

Pragmatism and common sense are needed to reduce some of the disadvantages of livestock farming and maximise its functions.

As small-scale farmers, livestock is not our core business. The various aspects of animal husbandry are well described in existing literature. We refer the reader there, and will not go into technical details. In these next pages, we will simply mention a few ways of thinking about the different roles of animals in a permaculture microfarm setting.

Maximising the Functions Performed by Animals

Animals often have one main function, such as providing eggs or milk along with several additional functions. Industrial agriculture focuses on the main function and neglects the rest. Permaculture thinking invites us to analyse all potential uses of animals in depth, to get the most out of each one. This is only possible on a small farm.

We have already discussed the advantages of integrating a chicken coop into our greenhouse (see volume one and two). We benefit from eggs, compost, warmth, also the work of chickens on weed control and slugs that, in turn, reduces food costs for our chickens.

A draught horse offers us its working strength, but it fulfils many other functions since our daughters ride as a hobby. He turns grass into manure, contributes to the positive image of the farm, serves as a support for educational workshops, and makes a fine companion.

Limiting Dependence on the Outside World

Buying feed for our animals makes us feel uneasy. These feeds have an ecological cost, especially cereals. We prefer to use the resources of the farm. We are searching for ways to be self-sufficient in feed, which has led us to make hay ourselves, to plant fodder hedges, to grow maize, sunflowers and fodder cabbage, to inoculate composts with good bacteria so that the hens can partially feed on them, to give a few examples. We also limit the number of animals to match the capacity of the farm. This is a low-input, low-labour and low-productivity approach, but it still maintains the ancillary benefits of having farm animals.

Hardy Breeds

We do not want to make expensive investments in buildings where animals would spend the winter. The construction of a barn leads to a chain of dependencies: to make the milkings profitable requires intensive production.

It takes copious amounts of feed to support it, so you have to produce even more. Next, a building to store the feed may be needed and then a tractor with a lift to transport feed, and another building to house the tractor. The operation will also need a fuel tank, a separate shelter for the oil tank. European standards must be met as well. This logic of perpetual growth has ruined the farming industry. Leaving animals in the wilderness alleviates these constraints. They live in better conditions in the wild.

We only have animals on the farm that can cope with spending the whole year outside. We have built them small shelters for the winter, where they can protect themselves from the weather if they wish, but they often prefer to stay outside, whatever the weather conditions. Our Merens horse comes from the Pyrenees where its cousins live in a semi-wild state; our donkey is well adapted to the Norman climate as it is a Cotentin breed; our hens were born on the farm, generation after generation. They go about their business freely and we only provide them with supplementary food.

We are attempting to foster a local breed of 'Bec Hellouin sheep'. Initially, we had sheep from the island of Ouessant. They are extremely hardy, living outside all year round, lambing without assistance, asking for grass because they do not like hay. But they have the disadvantage of being the smallest sheep in the world and usually only give birth to one lamb a year. We crossed them with Île-de-France ewes, a large breed of meat sheep that is very productive but known to be demanding with upkeep. The result was sheep with traits of both breeds, very hardy, with delicious meat. Unfortunately, they kept the Ouessant characteristic of giving only one lamb per year. Next, we introduced a third breed, the Shetland, with intermediate characteristics, hardy but larger than the Ouessant and birthing two lambs per litter. One or two generations later, the reproduction rate has effectively risen to two lambs. At present, we select breeding ewes according to the criteria we are looking for including medium to large size, white colour with occasional red tones and good fertility. These sheep perfectly meet our expectations as they maintain our orchards without requiring any care, apart from one shearing, one hoof trimming and one worming treatment per year, and they give birth to their two lambs by themselves. They are also perfectly adapted to the climatic characteristics of our valley, as well as to what our grasslands offer as food. The meat of these sheep who have a healthy and natural life is quite exceptional, at least equivalent to that of the most renowned breeds.

In its quest for high yields, production agriculture maximises worries, but not necessarily profits. In the fifteen years that we've had sheep, they've acted independently and led largely independent lives. However, we visit them daily and keep an eye out for problems that are usually related to fencing, as we shall see later.

▲ *Living close to animals, from birth to death, facilitating their needs and reproduction gives our children great life lessons.*

Holistic Pasture Management

We were inspired by the work of Allan Savory, the originator of holistic management. He studied herds of large wild ruminants, such as buffalo and wildebeest, and found that they often live in large, densely packed herds. These herds, which may number in the thousands, are constantly on the move, seem to graze everything in their path, and leave a considerable amount of dung on the ground. Based on the behaviour of these wild animals, Savory found that herds of hundreds of cattle can graze a semi-deserted area despite a lack of vegetation, if they are moved weekly or even daily. Against all odds, this approach allows vegetation to recover quickly and diversify. This is one of the most effective ways of regreening deserts, and a profitable one at that, since it produces meat.

Holistic pasture management involves a series of small grasslands. Animals are concentrated in one section for a few days and that area is then left to rest for two months or more, depending on the growth cycle of the grass. This improves the condition of the grass, prevents soil compaction and desertification by eliminating animal damage to trees, it reduces the presence of parasites and allows more animals to be kept per hectare. We have adopted this system on 5 hectares with twelve different plots, in addition to gardens and a forest garden. To increase the benefits, the plots are all planted with fruit trees and surrounded by fodder hedges.

▼ *Fénoua and her tame chick. We could never bring ourselves to tell her that one of our staff's dogs put an end to her.*

The Close Association of Animals and Plants

We find it very satisfying to explore small agriculture. It is free from large machines and gives farming a human face. We are trying to recreate bonds between plants, animals and humans. We bring animals closer to cultivated plots of land like the forest garden, that can

▲ *Lila and a duckling. One day, one of our daughters came back from the garden and said, "Daddy, I went down to the pond and a duck smiled at me, so I smiled back at him!"*

occasionally be used as pasture, or the fields after harvest or at the end of a green manure cycle. This way, all plant production is put to good use.

We arrange plot runs so animals pass by a set of plants that are beneficial such as comfrey, nettles (once mown and dried, all of our animals will eat nettles, particularly useful for nursing females or in case of food shortage), dandelions, lupins, alfalfa, wormwood or, in the same family, tansy (excellent vermifuges), mugwort or *Artemisia annua*, amaranth, to name a few. Of course, it is important to check that each plant is suitable and healthy for your animals. Between book knowledge and observation, you will know the best plants for each type of animal and be able to create runs offering a natural, healthy and adapted diet. These runs are particularly effective and easy to set up for chickens.

Reducing the scale of the system will increase the number of interactions and efficiency of the system, again, the permaculture principle of small solutions.

Fencing

Having multiple plots requires excellent fences. Every animal's dream is to get into the next plot, where the grass is necessarily greener or the crops more appetising. Despite our best efforts, once or twice a year we wake up in the morning to find the horse or a sheep in a garden bed. As for the hens in the greenhouse, there are always a few habitual delinquents who make their way into the crops.

Fences can never be too strong. We prefer good fixed fencing - sheep wire with five rows of barbed wire. Electric fencing is quicker to install but less reliable and needs to be moved regularly. Holistic graziers commonly use electric wire fencing. Putting up good fences is a big investment, and creating a good plot run is a lot of work, but in both cases your daily life is improved in the long run.

▼ *According to John Seymour, one of our teachers, there are many advantages to having a cow, even on a very small farm.*

Feeding

Estimate how many animals you can feed on your farm, including newborn estimates in the coming year. Farm literature provides benchmarks like one cow or one horse, or ten sheep per hectare that will need to be adjusted according to your land and your practices.

A basic feeding portion covers the animal's vital needs, additional feed will give it the energy to provide what is expected of it in addition to basic survival such as labour, egg laying, milk production, etc. If production declines or is not up to scratch, increase the ration. On the other hand, a resting horse or a cow that is not breastfeeding just needs its basic ration. Too much feed can be harmful.

A Large Animal on a Microfarm?

Having a very small area makes it challenging to keep one or more farm animals, but since animals prefer to live in company, it would be cruel to have only one animal. Nevertheless, we believe - and this is partly at odds with the above - that it may be wise to keep a small draught horse, or a donkey, or a few sheep, or even a small cow, even if you only have a small area of less than one hectare. This means buying hay and straw from outside the farm to supplement their feed and for bedding, which can be costly. The presence of animals will provide you with manure to fertilise your garden. Not to mention the pleasure they bring. The downside is that you will need to be present on site daily to provide the extra food.

Choosing Animals

The way of ecoculture leads us to favour local, rustic and versatile breeds, rather than modern, specialised breeds that are ultra-productive but fragile and require more care and food.

By choosing to raise an animal, the farmer becomes responsible for its welfare. It is important to learn about the needs and habits of animals, not only from literature, but also by carefully observing the behaviour of the first generations of animals on the farm.

As mentioned in volume one, every animal has functions and needs which must be considered:

- o Food
- o Water
- o Shelter
- o Fencing
- o Regular care: shearing, trimming, deworming
- o Exceptional care: diseases
- o Reproduction and farrowing
- o Hygiene
- o Protection
- o Legislation (e.g. declaration for sheep).

The following notes are not a guide to breeding, just some food for thought.

Chickens

If you have to choose just one type of animal, it should be chickens. These modest birds are content with little and give much. Their diet is largely made up of what they gather in the field. Kitchen scraps and a few handfuls of grain complete the diet. Access to the compost pile gives them a source of protein to replace grain – according to some experts, compost contributes 10% of their daily ration.

In Asia, Perrine observed a clever system that we have adopted. Chickens need a good henhouse that is closed at night because of foxes. What they appreciate most of all is being able to spend the night perched. This makes them feel safe. Their droppings accumulate under the perch. This compost heap in the making can be regularly enriched with carbonaceous material, possibly with microorganisms, and aerated. If it is exposed to sunlight for part of the day, it will give you high-quality compost.

Do not allow chickens access to market garden beds, except possibly in winter. While they do eat slug larvae, they cause too much damage by pecking at young leaves and scratching at mulch. Chickens can be kept under very light supervision, especially if you use feeders and an automatic gate to close the house at night.

Geese

Geese are elegant birds that feed mainly on grass. They have a role to play on a small farm. They are hardy and resilient and produce large eggs and succulent meat. If you have an effective mobile fencing system, they can maintain grassy areas such as driveways. As young birds, they can understand some instructions and will follow you from one plot to another. But their reputation as good guardians is not misplaced, especially if a gander is part of the flock. If your farm is open to the public, your design should prevent any unexpected pinching.

Ducks

Ducks love water. Some breeds require a pond to thrive, while others need little water other than for daily grooming and drinking, such as Indian runners. On our free-range farm, they cross the river and enter the gardens where they can cause damage. In addition, ducks tend to trample heavily around their shelter or pond, turning the ground there into a mess. Your design will need to take all of this into account, even if you are only keeping ducks temporarily.

The Khaki-Campbell and Indian runner breeds are known to feed on animal protein in their youth, so they are particularly fond of slugs. Indian runners reduce the gastropod population. However, it is best to allow them access to the garden for only a few hours a day. Be careful, as they are a gregarious and timid breed. You will have to adopt them in a herd and let them get used to you. During their first six weeks on the farm we contain them in a hermetically sealed area like a chicken tractor and visit them frequently so that they get used to us. The habit is formed from the very first days of our visit to signal our approach with a rallying cry, "Duck-duck!" Perrine even signalled to eggs in this way and the ducklings recognised her voice as soon as they hatched. When they are left

▲ *This beautiful illustration, taken from one of our old agriculture books, shows a rolling chicken coop. Is this the ancestor of the chicken tractors that permaculturists love?*

▼ *Geese and ducks in front of the dog house. Many visitors to our farm are deeply moved by the intimacy between animals and people on a small farm. Ours is a 'human-sized' farm, for everyone!*

▲ *These farm-bred Indian runner ducks have proven to be excellent slug predators.*

free in the garden, we can encourage them to follow us where there are many slug attacks and ask them to 'dig' under the mulch. Otherwise, fearful as they are, they would tend to stay in the vicinity of their habitat, a floating hut that is very effective in preventing night-time predation.

At eighteen months Indian runners begin to vary their diet. They can devour a large number of young shoots in a short time if given the opportunity. Vary the areas they graze, especially in winter when grass is not growing, and contain them in an enclosed area with water and a daily food supply. If you can't set it up this way, consider introducing them only in case of heavy slug infestation, and limit the duration to a few months. But that would be a pity in the end, because, apart from slug control, the Indian runner is excellent for laying eggs. Philippe Desbrosses describes egg laying daily throughout the year.

▼ *The hardy Ouessant ram, the smallest breed of sheep in the world.*

Turkeys, Guinea Fowl

These animals are more fragile and do not cope well with a semi-autonomous diet. If you are motivated to devote time to the barnyard, they deserve to be welcomed, if only for their beauty. Again, with time, patience and attention, these animals can become more friendly than you might think.

Rabbits

There are many superb breeds of meat rabbits, but their breeding is rare on permaculture microfarms as one cannot envy the fate of these unfortunate mammals, locked in cramped cages until their last day. In Cuba, however, every farm that claims to follow permaculture principles has its own rabbit cage farm, where the droppings are collected for compost and even dried to be made into rabbit food. Killing such a beautiful animal and preparing it is an ordeal that the prospect of a good meal does not justify. Rabbits also require daily care.

Equidae

Horses, donkeys and ponies have their place on a small farm, provided you have an interest in them. We will talk about them in the next chapter.

Sheep

Sheep are peaceful and discreet animals that are easy to keep. They have no equal in maintaining the meadows. Beyond the farmyard, they are probably the best possible animal for a microfarm.

Goats

Goats are terribly endearing and consume more woody vegetation than sheep. However, they are absolutely formidable because they are very difficult to confine. Good fencing is essential. Our first goats seemed to have supernatural abilities to cross obstacles. I sometimes locked them carefully in the henhouse and a few minutes later they trotted around the yard free again as if nothing had happened. We would like to have goats for milk production, once we have managed to reduce the other tasks that the farm requires.

Cows

We have no experience with cows, but we have often thought of taking in a small Jersey cow, so beautiful with big doe eyes, and particularly rich milk. From a self-sufficiency perspective, introducing a cow to a small farm makes sense if you process the milk yourself, into butter, cream and cheese, and feed the whey to the farmyard residents or a pig. Cows become depressed if left alone so get at least two. What made us give up on the idea is that we think milk is harmful to human health, even if home-produced raw milk is much better than commercial milk, and the burdensome constraint of having to milk once or twice a day, every day.

Pigs

Pigs are wonderful animals, closely associated with humans almost everywhere in the world, including the poorest countries. They eat

◀ *Shanti feeds our 'Bec Hellouin sheep', whose colourful coat reflects the diversity of their origins.*

absolutely everything and thus make use of the various and sundry waste products that a family or farm produces. They can also turn over land, making it easier to cultivate. There are small pigs, which are probably easier to mobilise in a system of rotating pastures. Some permaculturists have been experimenting for a few years with raising pigs in a quasi-autonomous way, especially under oak trees.

Combining Several Animal Species

There are many advantages to combining several species of animals in the same space. They do not eat exactly the same thing and graze grass at different heights, which makes better use of it. When parasites specific to one species are ingested by another, they die. We combine all of our animals together in the same grassland.

The Health of the Animals

If they are well fed and enjoy a healthy life in the open air, diseases in farm animals are very rare. A diversity of plants in their diet contributes to this. Many plants, especially those in hedgerows, have medicinal properties and animals know how to use them.

To end this chapter on an exotic note, I was told about an ethnic group in Asia that lives in houses on stilts. The humans defecate above the pigs' yard, which feed on the excrement. The pigs' excrement falls into a pond and feeds farmed fish. Ducks also help to complete the cycles by feeding on the aquatic plants fertilised by the excrement of all contributors. The pond water irrigates the crops. This is, however, a permaculture design model that we have not tested!

We thought it would be a good idea to mention the presence of animals, even though the main purpose of a microfarm is rarely to raise livestock. Animals contribute to the coherence of the farm, but don't provide the farmer with a livelihood because commercial breeding must be practised on a larger scale to be profitable. A neighbouring shepherd, who raises splendid sheep on 100 hectares, earns less money than our 1,000 square metres of vegetable crops. Animals are present in a microfarm rather for self-sufficiency and ecological efficiency. The work and costs involved are far from negligible, even from the perspective described in this chapter. All this has to be taken into consideration in the design of the farm.

In the end, the beauty of the animals and the pleasure of living with them will probably tip the balance in their favour.

▼ *Pigs and sheep enjoy fodder wood that has just been given to them, during a dry summer when grass was lacking.*

◄ *When I was a teenager, I dreamed of learning to ride, but my family couldn't afford lessons. The farm made this dream come true, and much more! As we close this chapter, Winik, our working companion for the past thirteen years, is retiring peacefully to our pastures and we are awaiting the arrival of a Comtoise mare to succeed him.*

7

DRAUGHT ANIMALS

I wasn't really cut out for the current era. Everyone is rushing around, but no one is going anywhere. People don't live, they just exist... Heavy technology and light minded people have destroyed almost everything that nature had to offer.

Vernon Cooper, Lumbee Native American[30]

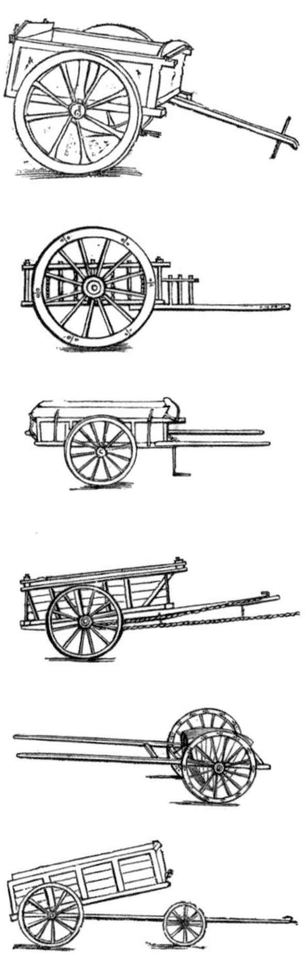

▲ *Before mechanisation, there were carts adapted to each use. Today, it has become difficult to find suitable equipment – old carts are usually very heavy and modern ones are ugly. Young entrepreneurs with good skills are needed to fill the void.*

Opting for animal traction rather than mechanisation changes a lot of things in market gardening. We made this choice from the very beginning of our farm and have always kept vegetable plots cultivated with the help of our horse, even after we discovered permanent bed farming.

Using a draught animal is not a return to the past, it is a trend on the rise - an ecologically coherent market gardening operation without the use of fossil fuels.

The following figures are surprising:

- o There are about 1.3 billion farmers in the world, or 43% of the working population.
- o 28 million tractors are currently in use, but only 2% of farmers use them. Sub-Saharan Africa has only 200,000.
- o 250 million draught animals ease the workload of 20% of the agricultural workforce.
- o 1 billion farmers work entirely by hand.[31]

Mechanisation is used by only a tiny proportion of farmers.

Animal Traction or Permanent Bed Farming?

For us, it's not quite this simple, as each approach has its advantages and disadvantages.

In our experience, permanent bed cultivation is the best in terms of ecological consistency, productivity, workload reduction and efficiency.

Animal traction requires a lot of skills; it disturbs the life of the soil and leaves it bare, it is not intensive because the horse has to walk between vegetable rows and is not very productive per unit of surface. Although we are now thinking about adapting the principles of conservation agriculture to animal traction, including no-till and sowing under plant cover, we have not yet found the right equipment.

We have decided to keep animal traction for some vegetables that are not well suited to raised bed cultivation. Ridging and harvesting potatoes destroys the bed. Squash requires a lot of surface area and grows just as well on a flat surface. Why bother making ridges if they have to be destroyed or if there is no reason to do so?

Animal traction also seems to us to be well suited to the production of common vegetables over large areas, storage vegetables with a long growing cycle and high demand like leeks, onions, carrots, and cabbage. A horse makes it possible to prepare a field relatively quickly and weeding easier. A large area of permanent raised beds would be needed to accommodate these crops.

Combining the two approaches is complementary as most of the production is carried out on permanent beds, with animal traction as a complement for field vegetables. We also use animal traction for experimental cereal crops and to transport heavy loads.

A draught animal on the farm also has all the advantages described in the previous chapter.

Modern Animal Traction: The American School and the French School

There are different trends and preferences among animal traction enthusiasts.

Some are adept at using old equipment, still found cheap at flea markets. That's how

"The evolving technology [of animal traction equipment] is ultimately less the implementation of simple and efficient mechanical means than a new form of social justice. To give the poor the means to rise by themselves, with their work, without handouts, without begging, without being dependent on anyone, and thus to become the equals of the rich in dignity, is indeed an informal act, implicitly contained in the assembly of a few steel bars."

Jean Nolle, 1989[32]

we started. However, it must be recognised that the crops of the past were ploughed, and those heavy tools required powerful horses. All these limiting factors suggest that a return to the past is not justified other than for historical reasons.

In the United States, animal traction is a living tradition, thanks in particular to the presence of the Amish, who have remained faithful to horses for religious reasons. Some equipment, such as horse-drawn manure spreaders, is still manufactured. The choice of equipment available is much wider in North America than in Europe. Animal traction is currently experiencing a revival with the publication of quality books and dedicated events.

This living horse culture has certainly preserved a heritage, but it remains rooted in a tradition of heavy tools that require two or three horses harnessed together. This is also true for the most modern equipment, which is sometimes equipped with a 3-point linkage, a power take-off or an auxiliary combustion engine. The price of this equipment is high. I am not sure that these horses in pairs or threes are more efficient than a single animal hitched to lighter equipment, such as we use at Bec Hellouin Farm. Each horse needs to be fed, cared for and shod. The burden of maintaining and harnessing several animals weighs heavily on a small farm.

In France, enthusiasts, often grouped in dynamic associations such as Hippotese or Trait d'avenir, have been researching and improving equipment for decades. Since 2006, we have been using the equipment designed by farmer-engineer Jean Nolle, a former cart driver, a visionary and pragmatic humanist, who dedicated forty years of his life to creating animal traction equipment for poor small farmers in developing countries. Jean Nolle carried out 144 missions in 70 countries and designed over 200 pieces of equipment. He has constantly sought to simplify his tools and make them more versatile. The result is an exceptional range that is very simple, very solid and effective. It is manufactured and marketed by the PROMMATA Association.[33] The low cost of these tools, their versatility and the ease with which a light animal can be harnessed have contributed greatly to the development of animal-drawn market gardening in France. We will come back to this later.

▸ *For want of a contemporary equivalent, we still use old equipment, like these hay rakes: durable and so economical.*

What can be Achieved with Animal Traction?

Almost all cultivation operations, for all types of crops, have historically been accomplished with animal traction. However, in the current context, in the West, the use of animals is still limited to market gardening, skidding logs in the forest, carrying out maintenance work, mowing, creation of waterways and, to a lesser extent, transport. Hardly anyone except the Amish are growing field crops with animal traction.

Animal traction is well suited to market gardening because animals compact soil less and do not create a ploughing surface. They do not require high investment and require modest installation costs without recourse to loans, which is in line with a desire for autonomy and respect for living beings that motivates people who make this choice.

Which Animal to Choose

Many animals can be harnessed; even dogs and goats used to be harnessed for small loads. The modern tools mentioned above do not require much power. They can be pulled by a donkey or a small horse, or even a pony. These three animals are used at Bec Hellouin.

We will not cover powerful but slow oxen and cows, which used to be very common. They are rarely harnessed nowadays.

The farmers of the southern hemisphere use all sorts of draught animals including zebus, camels, and elephants.

Draught Horses

There are many breeds of draught horses. In Western Europe, from the mid-19th century onwards, most breeds evolved into heavier horses that were powerful enough to pull the new machines that were appearing in the wake of the industrial revolution. They paved the way for the arrival of the first tractors.

Nowadays, despite the beauty of these splendid animals, I am not sure that their presence is justified on a vegetable farm. We had a large Norman Cob at Bec Hellouin Farm, and then a gigantic snow-white Boulonnais, whose hooves were as wide as plates, but these powerful animals can be tricky to handle and potentially dangerous. They cannot be put in the hands of novices and need more space to work between the rows of vegetables. This decreases the area of productivity. We finally opted for a Merens horse named Winik, who has been a faithful companion for thirteen years. He is the size of a small horse, weighs about 450kg, and powerful enough to pull a cart full of manure. Winik can be ridden and is ideal for market gardening operations. He is very inexpensive to maintain, and no longer shod since he rarely walks on pavement. A horse can go twice a week on the road without being shod. As a hardy breed, he can spend the year outside and has hardly ever seen a veterinarian since he came to the farm. A small horse like ours is an excellent compromise.

Compared to a donkey, the choice of horse requires more skill and care, but it offers greater versatility and strength. A horse is more fun.

Donkeys

The donkey is a very intelligent animal. Alice, our donkey, belongs to the Cotentin breed of donkeys, which were known for carrying milk cans back to the farm after milking in the fields. She can open any fence and even untie a knot or open a carabiner.

Donkeys are strong, hardy, and sober animals. They require less grass space for food than a horse, but do not like humidity and must have free access to a shelter in winter. They are gentle and better for beginners than a horse. A donkey is a very good companion for a microfarm.

The moderate power of a donkey is sometimes a limiting factor for some market gardening work and many donkey owners harness them in pairs when necessary.

▲ *We highly recommend harnessing a Shetland pony on a small farm. Requiring little in the way of food and care, a pony is a true companion, easy to harness, safer than a horse.*

Mules are more powerful, but rare these days.

Ponies

I don't know of any other farm that harnesses a pony for market gardening. And yet, many small farms have one. We were lucky enough to have a Shetland pony named Harmonie who belonged to our daughter Lila. For Christmas, we gave her a harness. To our surprise, although she had never been trained, Harmonie behaved from the very first moment as if she had been doing this all her life. She pulled the market gardening tools with all her strength and made it possible to accomplish real work, less quickly than the donkey or the horse, but real work nonetheless. She was obviously enjoying it. We made her a small cart that tilts back to unload. Harmonie valiantly carried loads of several hundred kilos from one end of the farm to the other. We wondered where she got this instinct to pull.

▼ *Teddy and Harmonie work on making a warm layer using the small cart we made from a pony carriage.*

We learned from some research that the small size of Shetland ponies comes from the fact that they were selected to pull wagons in coal mines.

Our farm is accredited as an educational farm, and Harmonie has enabled us to run wonderful educational workshops for school groups.

A Shetland pony easily weighs 200kg, which gives it far greater strength than a human. Its small size means that it can be fitted with a simple harness in no time. A pony is always available when help is needed. While it takes fifteen to twenty minutes to prepare and harness a horse, a few minutes are enough for a pony. The price of a pony remains reasonable and it costs almost nothing to maintain. Harmonie was bought for 200 euros. We have wonderful memories of this little collaborator who left us last winter. She was part of the family.

Good Training: An Essential Prerequisite

▼ *The choice of animal traction has never been questioned on our farm, even if the work is physical. On the contrary, we have acquired new equipment to carry out a wider array of tasks.*

To successfully start a farm with animal traction, it is not enough to have an excellent four-legged friend, you have to be very well trained yourself. I had several different training courses and was used to riding horses. Despite this, my early years were punctuated by clumsiness and a few setbacks. Driving is a complex and potentially dangerous art, especially if you go on the road. With traffic, a poorly controlled animal can cause an accident. However, it is more difficult to control a harnessed horse than a ridden one and not all draught animals are docile.

Training includes knowledge of horses and their needs, harnesses, equipment and a lot of practice to be able to deal with all situations. I experience many of the sensations of sailing while learning to drive horses.

You can train in specialised schools. Associations that teach animal traction also offer introductory courses. A training course with an experienced market gardener is also advisable.

Choosing an Animal

When you are a beginner and you are passionate, you want to buy a very beautiful animal. This led me to our first two horses, which were certainly splendid, but much too fiery and powerful for the beginner that I was. When the young Boulonnais, weighing a ton, started to do sheepish jumps in front of the tool, I didn't know how to react.

A proverb says: "Young driver, old horse". It is wise to choose an animal that may not make the front page of a trade magazine, but will wisely do what is expected of it. Winik was fourteen years old when we acquired him and just finishing a sporting career. Putting him to work in the fields required a long and patient period of adaptation. A mounted or harnessed horse is not specifically prepared for the prolonged and precise efforts required in market gardening. When the driver and the animal are both beginners, things become more complicated. Good training is essential.

Once you have been trained and have chosen a horse or a donkey to be your working companion for many years, take the time to look for the rare pearl – a calm but willing and friendly animal with whom you have a good feeling. Avoid animals that are too timid and put you at risk of an accident at the slightest unexpected stimulus.

Training a Draught Animal

It is very rare to find a well-trained draught animal. Market gardeners tend to keep their good collaborators.

Working the land is different from harnessing for pulling a vehicle. However, an animal that is already well trained for driving should have no problem pulling farm tools.

◄ *Sometimes we harness our donkey and horse together for work requiring power.*

Ideally, you should take your animal to an experienced market gardener and work together for a few days.

If you are a real beginner, we advise you not to undertake self-training. An interest in ethology allows you to understand equine psychology in detail and achieve a true partnership with your animal.

Equines are fearful because they have no other defence than to flee. In the wild, horses live in bands led by the dominant mare. By your calmness and strength of conviction, show your horse or donkey that you are the dominant animal. They will not take your word for it; they will test you. Do not give in. Once they recognise you as 'the boss', they will be reassured by your simple presence and willing to do what you ask.

At age three, a young horse can be harnessed. Take it easy! Start with light loads, such as a tyre sliding on the ground, and gradually increase the effort. Be careful not to put them off by asking too much.

Pulling a tool that is embedded in the ground requires great effort from the animal, who needs to be encouraged and rewarded, not beaten. Be patient. A horse is like an athlete; it takes time to build up strong muscles. After winter, your horse needs to be trained gradually over several weeks before it can perform to its full potential.

Given its body mass, a horse also needs time to warm up at the beginning of a session.

Feed

A working horse must be well fed; grass alone is not enough to provide the necessary calories. A horse's feed ration should not be changed suddenly; it should be increased gradually. It is the previous day's feed that provides the energy for the next day. Literature on this topic provides guidelines for calculating rations.

How Much can a Horse Work?

In the large farms of the past, horses that were specialised in field work were the strongest. They were trained and fed like top athletes and routinely had a long session in the morning and another in the afternoon. This was a hard life for both people and animals. The career of these workhorses lasted only a few years, after which they were considered fit only for the slaughterhouse.

The farmers who worked with their horses on a daily basis reached a level of teamwork that we can hardly imagine. Farmers could direct their horse from a distance by voice. Today, it is difficult to have such intimacy because the working sessions are more spaced out. Ideally, one should work at least twice a week so that the animal is always in top form. This is possible on a farm that has chosen to do everything with animal traction, but at Bec Hellouin we are far from this since permanent bed cultivation is our main activity.

Choosing animal traction is not something you can improvise; good training is essential. It allows you to choose an animal with good character, find suitable equipment, and above all, to work safely.

▼ *The equipment manufactured by the PROMMATA Association gives us complete satisfaction. They illustrate what can be done when you combine the knowledge of the past with that of today. In this photo, I have moved the spreader bar that connects the horse to the tool holder to one side so that Winik can walk between the ridges while the tool is working on it – the objective being, of course, that he does not trample vegetables. Note the very simple harnessing, sufficient to pull the tool carrier.*

Harnessing a Draught Animal

Once you have chosen your animal, you have to choose the type of equipment.

Collar or Tinker

A collar is a heavy piece of harness made of leather and metal that girdles the neck and allows the horse to pull in the most ergonomic way.

The breast band is a lighter piece of tack that goes under the neck and around the chest.

For agricultural work, it is best to opt for a collar that allows the animal to develop about 20% more power. The tinker is best used for leisure driving. The collar is more ergonomic and comfortable.

The Harness

The harness must be perfectly adapted to the animal's physique, solid and well maintained so that it does not give way during effort. There are many models; for the market gardener, simplicity is often synonymous with efficiency.

A collar and two lines: the simplest solution. The collar can be connected to the equipment to be pulled by two lines, which can be two ropes, two leather bands, or possibly two chains. One end of each line is attached to the collar, the other to the spreader bar, i.e. the piece of metal that connects the animal to the equipment to be pulled. A collar and two ropes are sufficient to harness an animal. We use this simple set-up for both donkey and pony.

A collar, two ropes and a sling are almost as simple. Harnessing an animal can become a bit time-consuming if you install a full set of equipment for each session. When you work with your horse and the days are long, you don't want to add an extra hour to 'trim' and 'untrim' your animal and then wash the harness. A simple harness made by our saddler is perfect. This model has proved to be well suited to farm work for over ten years.

The collar, or in our case the prong collar, is connected to the spreader bar by two lines. To prevent the horse from passing a hind leg over a line when it turns around, we have simply connected the collar to the tail with a leather band and put two elastic bands on the croup to hold the lines.

Another practical improvement deals with the traces that lie on the ground and risk being trampled and dirtied when the horse is not harnessed. We made the traces short enough so that they do not touch the ground. Two short chains on the spreader bar make up for the missing length so that the horse is at the right distance.

This extremely simple model is perfectly suitable as long as the horse is not hitched to a cart or machine that needs to be slowed down when going downhill. If this is the case, a full harness should be fitted.

▲ *Winik, a former harness race horse, won several trophies. It took us a long, long time to learn together to work slowly and precisely.*

There are a variety of full harness designs for pulling carts. In addition to the elements already described:

- A braking device that wraps a leather band around the rump. The animal controls the towed vehicle by braking with its body against the strap.
- If the cart has shafts to pull, a small saddle is placed on the animal's back. Two leather and metal loops on the saddle hold the shafts in place.

A complete harness consists of parts for pulling, braking and attaching cart handles.

Net and reins. Whichever harness is chosen, a snaffle is placed on the horse's head so the horse can be steered by the bit. The reins are connected to the bit and serve as a connection between driver and animal. The snaffle is usually fitted with blinders made with pieces of leather. They limit the horse's peripheral vision, so that it is not frightened by the tool it is pulling, that big thing that insists on rolling so close to its hindquarters. If the animal is placid, some of this equipment isn't necessary. Our little Harmonie was simply led with her lead rope and halter. As you can see, when you are breaking in, there are a lot of pieces of leather that you have to get to know, put on and adjust so that the animal works comfortably.

Soil Working Tools

Here is a quick overview of the most common tools used in animal traction. They are the same as those used with a tractor, only smaller and lighter. The implements are presented in the order in which they are most often used during successive cultivation operations: soil preparation, cultivation, crop maintenance and harvesting.

The Plough

The plough has long been the farmer's tool par excellence. Metal ploughshares and mouldboards that turn over soil became widespread in France at the end of the 19th century. It is still fairly easy to find old Brabant ploughs that have two ploughshares on top of each other, and can switch from one to the other at the end of a row to make all the furrows flow in the same direction. We have two such ploughs on the farm, but they require a very strong horse, as ploughing requires so much power. We do not use it very often.

The Mouldboard Plough

An alternative to deep ploughing, the mouldboard plough turns over grassland or green manure by working on a vertical plane without mixing soil horizons. It consists of three small ploughshares that pass under the surface of the soil to break it up.

Cultivator-Harrow (Vibrocultivator)

This tool is a kind of harrow with long, curved tines that give it elasticity. The tines penetrate the soil and break it up without mixing layers. It vibrates as it moves and this helps break up clods. The vibrocultivator lacks the power to attack a well-established meadow or green manure, so it is used after ploughing.

The Harrow

A harrow consists of a metal frame with tines that scratch soil. They break up the clods, refine the soil and level it. After ploughing the harrow prepares the ground to receive crops. It can also be used to scarify grassland.

▼ *Ridging discs lift soil to make a ridge. Another tool, triangular in shape, makes a small furrow at the top of the ridge that facilitates transplanting. The horse walks in the furrow that separates the ridges.*

▲ *The harrow is a fine tool, capable of weeding certain crops with a single pass if the right time is chosen to intervene.*

▼ *After passing the harrow over the top of the ridges, the weeding share can be used to work between ridges. Changing tools is quick and easy, but good control of the horse and the equipment is essential, as the slightest deviation would cause these onions to uproot.*

The Subsoiler

If soil is very compacted, or to break up a ploughing sole, a subsoiler can be used. It has a strong blade that plunges quite deeply into the soil.

The Ridger

This tool consists of two metal discs inclined on their axis. As they move forward, they lift the soil and form a ridge. This tool can only be used on well-prepared soil. The current model offered by PROMMATA has a tine behind the discs that digs a furrow on top of the ridge. This is very practical for planting crops while remaining perfectly centred on the bed to avoid damage during subsequent weeding.

The Harrow-Rower

This type of harrow is equipped with numerous flexible tines, sharply inclined towards the rear to weed well-established plants. The tines pass over the crop and lay down the young plants without damaging them and scratch the soil just enough to remove new weeds. The operation is a little stressful, but effective.

Furrow Cutter

This tool is used to weed furrows between ridges. It consists of a ploughshare that works flat on the ground. Blade width can be adjusted to suit different spacings and crops.

Weeding Fingers

Used along the edge of crops, this tool has two crowns that rotate when moving forward. The upper ring is made of a soft plastic to weed plant tops. The lower ring has metal fingers to weed the top of the ridge. When properly adjusted, these tools are very effective.

Pig's Ears and Crow's Feet

Tines and blades of various shapes used to weed the furrow and the sides of the ridge.

Potato Harvester

This tool is used to open ridges to extract tubers. It consists of a ploughshare that plunges into the soil with a screening surface attached that lets the soil fall through and captures the potatoes.

Tool Carrier

Many animal traction systems are designed around a versatile tool carrier that can support a range of equipment, such as those described above.

PROMMATA Tools

The PROMMATA Association has taken over and further developed the tools created by Jean Nolle during his long career. These tools are designed by farmers for farmers. They have given us total satisfaction for thirteen years now. Readers wishing to learn more can refer to the association's website.[34]

On the farm we use the tool holder called Kassine, of which more than 1,600 have been made to date. It has several different attachments that make it wonderfully versatile and comfortable to use.

The Kassine can tow all of the above tools and many more. Changing tools is simple thanks to the 'crochaxe' system developed by

Jean Nolle. The new Kassine model can carry three different tools, allowing several operations to be carried out in one set-up.

We also have a multi-tool machine called Le Mata, which allows the operator to work sitting down. It can hold the same tools as the Kassine but can work on two ridges at a time. The strength and reliability of these tools are such that they can last a lifetime. No programmed obsolescence!

The Ridge Cultivation System

The Prommata range of products performs almost all cultivation operations used in market gardening. The system is based on ridge cultivation.

A ridge is a temporary bed, utilised for the duration of a crop. It is about 25cm high and 50cm wide and the length of the plot. We worked with plots that were 40 metres long and found that this is too short for animal traction because it requires frequent turning around. Turning around takes time and is tiring because you have to carry the back of the tool and set it in the new direction. Our animal-drawn plots are 100 metres long.

Raised ridges guide the horse and make it less likely for the animal to trample crops that are raised up from ground height.

Ridges allow for easy weeding between rows and offer greater depth of topsoil for improved ventilation. In the row, manual weeding is required.

When the soil is decompacted and ridges formed, weed three or four times between rows with the horse, which is faster, and one pass by hand on the row, to keep the crops weed-free until harvest. For temporary ridges we are much less demanding on weed control than with permanent beds. Weeds can contribute to fertility as long as they do not dominate the crop.

The Disadvantages of Ridge Cropping

Low Density

The fact that we and the horse walk in the furrows means that the ridges have to be at least 70cm apart, ridge to ridge. A row of vegetables every 70cm is fine for potatoes or cabbage, but not for leeks, onions or carrots! Animal traction means that a large area of soil has to be prepared for a relatively low density of vegetables.

We try to get around this by levelling the top of the ridges so that several rows of vegetables or combined crops can be planted. The chapter on combinations (volume two) describes our attempts to do this.

▲ This spreader bar was moved to the side so that the horse can walk between rows. It easily detaches with a snap hook. Note, in this photo, the tool carrier is equipped with its two wheels, whereas it has only one to pass the weeding share, as illustrated in the photo below.

▼ For some of our trainees the discovery of animal traction was such a powerful experience that it led them to change jobs.

▲ *Nothing could be simpler than putting baskets on the animal's back – care must be taken to balance the loads, however. Harmonie was happy to take on all sorts of work, here the stripping of the plots above the Bec Hellouin Abbey.*

Working the Soil

Market gardening with animal traction is still based on regular tilling, a negative even if it is much less aggressive than mechanised work. When practising absolute respect for the soil, this is not at all satisfactory.

Bare Soil

Soil is often bare and subjected to the aggressions of sun, rain and wind.

Fertility Renewal

With both frequently tilled and bare soil, fertility is lost to bad practices and crop exports. The question of fertility renewal is acute. We look for answers to this question in several ways:

Green manure: in our animal-drawn plots, we only carry out one crop cycle per year. The land is then sown with green manure. Although this practice is satisfactory from the point of view of fertility, it is not technically sound because horses lack the power to take down well-established vegetation over a large area. Ploughing used to be the answer, but we are trying to avoid that. Cultivating a grassed or green manure plot with a trisoc (plough with three shares) and then a cultivator requires lots of passes spaced out a week. For a well-weeded plot, these passes will have to be repeated for about two months depending on the weather.

Mulching: an alternative to green manure is to fill the furrows with mulch, such as manure, once the crop is well established. This requires a large amount of mulch and is time-consuming. However, it seems to be the best solution as it benefits the current crop, reduces weeding and keeps soil fertile and loose for the next crop. This approach allows for two reasonable crop cycles per year.

Transport

In modern times, transporting loads by animal traction has been less studied than tillage tools. It is difficult to find a good cart nowadays in France, whereas in the United States and Canada they are still being made. It is so practical and pleasant to hitch your horse to a cart!

Of course, a trailer coupled to a tractor makes it possible to work without fatigue and much faster. But for me, harnessing my horse is anything but work, and I prefer to finish an hour later and have the feeling of an accomplished and fulfilling day.

On the other hand, when you calculate the real cost of combustion engines, cars, tractors, etc., including purchase, maintenance, repairs, fuel and insurance, you can see that you have to work a lot of hours to pay for all these costs. So, all in all, working slower with a horse that has almost no cost turns out to be a good solution.

We have been experimenting with load transport with a variety of horse-drawn vehicles:

- A Swiss army sled: it allowed us to haul the crops from the gardens to the shop when Perrine and I were alone on the farm. It was quick and fun: the baskets were put on the sled, the horse was fitted with the simplest of harnesses and off we went.
- A Canadian four-wheeled cart: versatile, it is used for all purposes. It weighs 450kg and loads 1 tonne of manure.
- A two-wheeled Swiss army cart: this tiny

vehicle, which weighs only 70kg, is used to go on very steep terrain above the abbey.
- A small pony cart that can also be operated by hand.

We plan to make a horse-drawn cart from the four-wheeled cart. And if Father Christmas really exists and reads this manual about sleds, we would like him to bring us a manure spreader one day! US-produced spreaders, from Millcreek, are distributed in France by H2C.[35] There are some very small ones, suitable for a small horse.

Baskets

We have different sized baskets that are very practical for carrying loads on an animal's back. They come from Morocco.

▲ *These baskets are made in Morocco from recycled tyres. Simplicity of means, almost zero cost, no recourse to fossil fuels: this is the approach we like!*

Happiness in the Meadow

Beyond all the advantages and disadvantages of the 'animal traction' option, there are the joys of a relationship with animals. I remember one morning, when everything was white with snow, riding through the forest to cut firewood, and another time, with two precious friends, we went to cut down logs to build a shelter, all with Inou, our faithful little dog, always the first to jump into the cart to ride along with me. The family outing to the abbey in a sleigh brings back happy memories. It felt like we were living in Tolstoy's time.

To conclude, I would like to share an experience that has deeply marked me. I hesitated to write this down and have not mentioned it before. If you think I'm crazy, don't say it too loud!

I had heard that some people have extraordinary communication with animals, but I had never explored this. And then one day... It was the first work session of the spring and Winik, who like me is getting older, was out of breath. We had almost finished restoring a market garden, but there was still a thick strip of grass flush alongside a row of fruit trees. We were walking side by side, when suddenly the trisoc got stuck in a root and Winik was stopped dead in his tracks. Carried away by my momentum, I passed the horse. Then a message, not words, but a message, was imprinted on me in a flash, saying very clearly: "You can count on me, I will do it." I turned and saw my old horse stand up on his hind legs and throw his weight forward. He pulled the root out and we set off again. I had never seen him throw himself forward with such determination. Winik's behaviour matched the message so perfectly and changed our relationship. I no longer saw my faithful companion in the same way. When he is afraid and stops (he is quite shy), I speak to him silently in my head and he carries on.

◄ *I have given all sorts of rational arguments in this chapter for animal traction, but it is obvious that our main motivation for working this way is because it's not work, we love it! And then sometimes, too rarely, we use our four-legged friends for a family outing, like this morning when we all went to the abbey by sled.*

XIV

HAY AND GRAINS

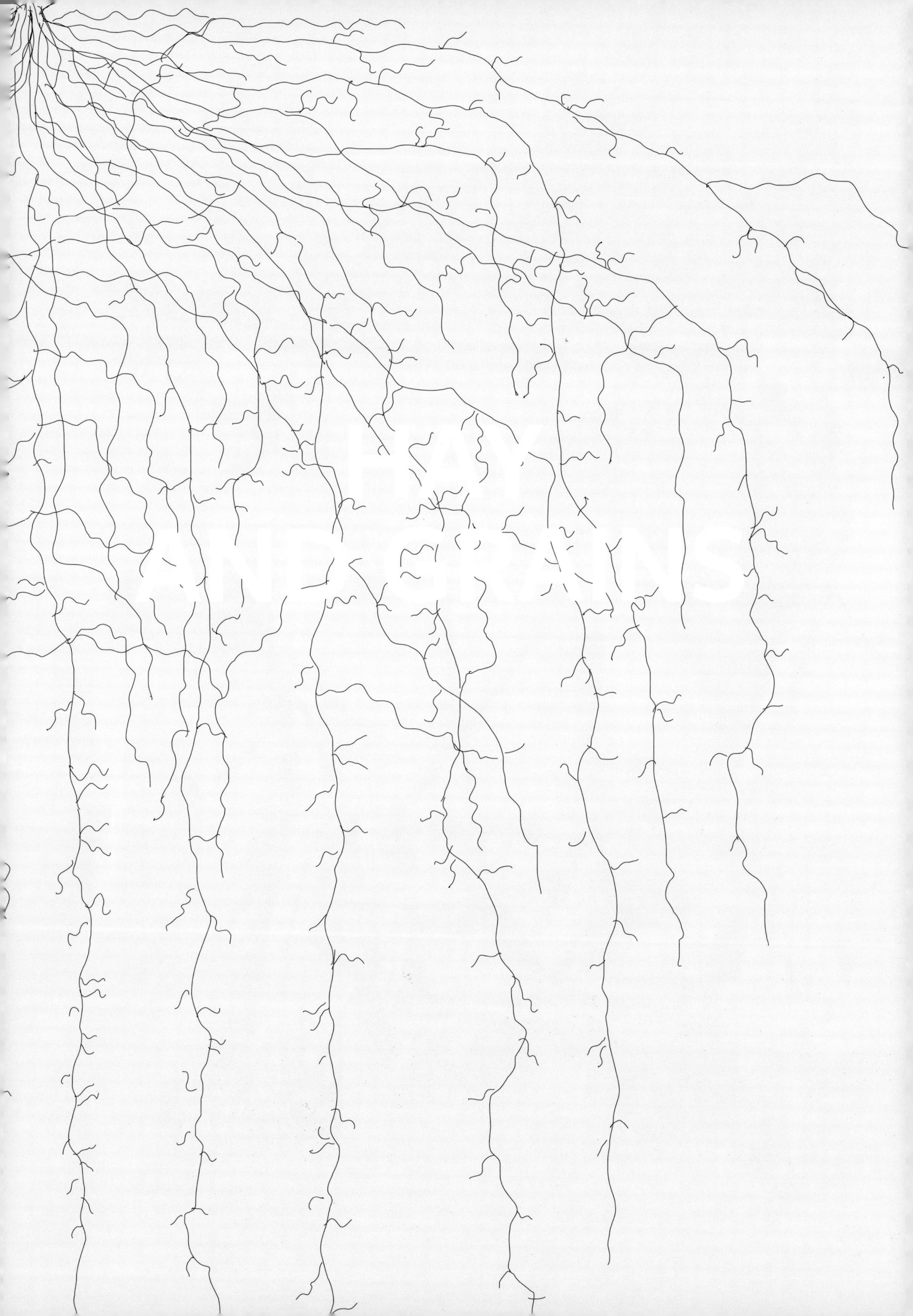

◄ *Making hay without petroleum? You can still find beautiful mechanical mowers with animal traction, like the one here, used by Sylvie Dévigne, a young woman who comes to mow our meadows. To pull such machines, you need a strong horse or a pair of horses. The permanent presence of such an animal is not really justified within a microfarm, especially as it needs a hectare of meadow to feed it. Hence the interest in using a service provider, like Sylvie. Animal traction professionals have excellent horses and the best equipment, as well as a high level of expertise.*

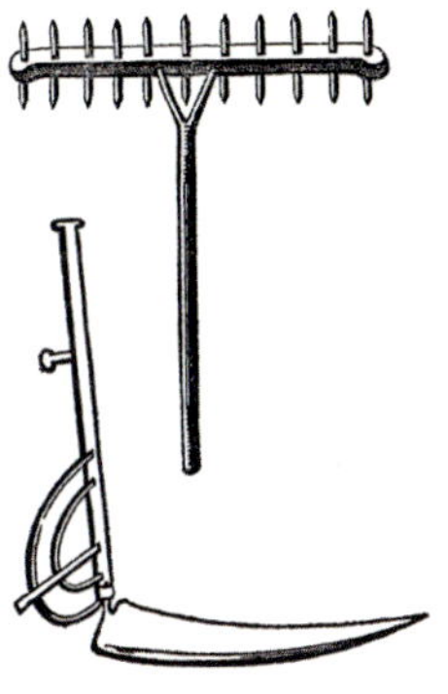

▼ *Mowing a meadow by hand is justified for small plots. When you have mastered the sharpening and handling of a scythe, making hay is a real pleasure. Under the June sun, the meadow is a soft and fragrant universe, a great place to immerse yourself.*

8

HAY

Sovereignty is something that spreads to wider and wider circles, starting from yourself... Sovereignty is not something that is granted to you... It is not a right that you receive from outside. It is something you carry inside.

Eddie Benton-Banai, Ojibwe Native American[36]

Having animals on a microfarm requires feeding them year-round, as naturally as possible, ideally from the farm's own resources. This is feasible: farmers in the past raised their animals without buying disastrous GMO soya from Brazil, the cultivation of which is contributing to the destruction of the Amazon.

A grassland system based on very good meadows surrounded by fodder hedges makes it possible to keep animals outside all year round, at least in our latitude. This is facilitated by the choice of hardy breeds. In England, dairy farms raise their livestock in the fields all year long. The animals are less productive, but much healthier than cows that are fed protein meal and end up prone to mastitis. Reduced operating costs allow for a decent net margin. Everyone wins: the cattle, and the farmers who have less work to do.

In winter, however, grass can run out. In France, growth stops between November and March. The hedges no longer have leaves to provide additional food. In this case, the food ration is supplemented with hay. Cutting grass in the summer, when it is full of nutrients, and storing it as hay for the winter redistributes the sun's energy at a time when it is lacking.

What is Hay?

Hay is dried grass. Mowing a meadow yields a varied mixture of grasses, legumes and other plants, depending on its composition. Hay contains many fine stems and leaves that give it a green hue and pleasant smell.

The best time to mow is when the grass is in flower or just after. This offers the highest concentration of energy and nutrients.

Hay should not be confused with straw, which is the stalk of cereals that are yellow in colour, thicker than hay, and contain more cellulose and waxes. The digestibility of hay is better.

It is possible to make two cuts of hay per year, but the second cut of hay is less abundant. We do not make a second cut to avoid exhausting our meadow, which receives no fertiliser other than animal excrement.

How to Mow a Meadow by Hand

When a meadow reaches its optimum stage, around the summer solstice in Normandy, we take advantage of a period of good, stable weather to mow. The aim is to mow and dry in as little time as possible, as rain or a storm could compromise the quality of the hay.

We will deal here only with the artisanal way of making hay. This non-mechanised

▲ *The ancients mastered the art of the haystack. Different approaches are possible. Stacks could be temporary, before hay was brought into a barn, such as the haystacks shown above, or more elaborate, allowing outdoor storage during the winter season.*

approach is well suited to a microfarm whose small area does not justify owning machinery. If you don't want to try this, you can use the services of a neighbouring farmer.

In our quest for self-sufficiency and petroleum-free farming, we asked ourselves how we could do without modern equipment. We chose a baler design that makes small rectangular bales weighing about 15kg – their small size makes them easy to handle and store, and therefore practical for a microfarm over balers that make large round bales weighing about 150kg, designed for farmers with mechanised lifting equipment.

For a long time we bought our hay from outside and then one day we took the plunge. The weather was so nice in June 2015 that I took out my scythe, sharpened it as best I could and went to mow the thick grass of one of our meadows. Two days later my back was sore, I was exhausted and the meadow seemed to grow wider and longer. I called our friend Emmanuel Oblin, the creator of the *Comptoir de la Faux*, which distributes the best European scythe.[37] Emmanuel came to the farm the next day, and with his usual kindness, he showed me how to make a scythe as sharp as a razor, one of the great secrets of successful mowing. The other secret is good posture: back straight, legs apart, the scythe well-adjusted according to one's physique, the blade placed on the ground and smooth movement of the whole upper body.

Emmanuel's advice had an unbelievable effect. The backache disappeared and I was able to move much faster. Our whole team got involved and the mowing became a joyful group effort.

The scythe and its sharpening will be described in more detail on p.116.

▼ *This hay rake can work in two positions. Open, as here, it allows you to make windrows, i.e. to group the hay in strips to facilitate collection.*

Mowing with a Horse

This first experience with the scythe, however, left us doubtful about the amount of time we could devote to making hay in the middle of June, when market gardening demands our full attention. In the following years, we called on the services of Sylvie Dévigne, a young woman who has courageously taken up animal traction and provides a mowing service. Sylvie and her mare come to mow our meadow, and do in half a day what would have taken us a week with a scythe.

More recently, a friend gave us an old horse-drawn mower. Our Merens horse is too weak to pull it because it requires a lot of power, as the back and forth movement of the blade is driven by the rotation of the wheels – what the Americans call *ground drive*. We tried it out by hitching it up behind our old tractor. The mower works perfectly, after perhaps forty years in the shelter of a barn. We are now self-sufficient, but I think we will continue to use Sylvie and her mare, which compacts the soil less and is, let's face it, much more pleasant than a tractor.

Drying

We have an old hay rake at the farm. This beautiful machine is used to aerate the cut hay by turning it over. It can also be used to make windrows, or long strips so that it can be picked up more easily when it is dry. Our horse Winik pulls it without too much effort.

Drying is an essential step for the quality and preservation of the hay. It reduces the water content of the grass from about 80% when it is standing, to 15-20%. When the weather is good and dry, the hay can be turned twice a day and drying is completed in about three days.

It is also possible, if you do not have a horse and a hay rake, to do this by hand. Our ancestors did this with a large wooden rake, which is preferable to a metal rake that has teeth that dig into the ground. A good part of the day is then spent stirring the long, fragrant swaths. Don't think of these operations as a chore: on the contrary, they offer the best of country life.

Making a Haystack

The dry hay is loaded into Winik's cart and brought back to the farm. As we do not have a barn, we have to store it outside. We learned how to make haystacks and our apprehension that we couldn't pack the hay without a machine soon disappeared: it is really compacted very tightly in a haystack.

I must admit that learning how to make a haystack was a great joy. I had never seen one before in my life and it is really beautiful.

In order to be stored outside for a long time, the hay has to be ventilated and kept free of moisture. There are different techniques. We opted for an approach that we discovered on an American website that gives us complete satisfaction. A wooden support is made out of four poles joined at the top, like a teepee. They carry a platform made of poles and installed about 30cm up from the ground. The poles are simply tied together with string. This structure can be used for three or four years. The hay is placed on the platform, so that the air passes underneath, protected from the humidity of the ground.

The hay is raised regularly by turning the stack around. It should be noted that the hay is loose and not bound in a sheaf, which would be impractical. One person climbs onto the pile and evens out the hay sent in by the other participants; in doing so, he or she packs it down considerably, so that it will be difficult to remove from the stack. It must be raised above the top of the poles and shaped to allow water to flow down the sides.

The Art of Haystacks

In paintings and photographs of the past, one can admire truly impressive haystacks. Old books show the degree of perfection achieved by farmers who did not have large enough barns to store their crops. Haystacks and grain sheaves could become quite large. There were different solutions to protect them from moisture. They were built on an earthen mound surrounded by a ditch, on a bed of sticks or on a raised platform with supports that could be topped by a stone cover or tin cover to prevent rodents from gaining access to the grain. The millstones often had a central mast to enable them to withstand strong winds and sometimes a central chimney for ventilation.

The cover was done by an experienced worker, in the same way as a thatched roof. Straw mats were held together with straw ropes that were braided by twisting the stems.

Before we made our first stacks, I was worried that we would never be able to master these unique skills that had been interrupted with the arrival of industrial agriculture. How to arrange the hay so that rainwater runs off it? Fortunately, we can now quickly make simple and efficient haystacks. Do you know how to shape the outer walls of a haystack so that they have a consistent appearance, with all the stalks facing the ground? All you have to do is smooth them with a fork. You comb the edge of the wheel, removing the stalks that are not well fixed inside the wheel, and in doing so, you direct all the stalks downwards. In this way, the rain will run off without penetrating the stack by more than a few centimetres.

The top of the stack is protected with a tarpaulin that is firmly attached to the wooden frame.

A haystack made in this way allows us to store enough food for the horse, donkey, pony and sheep for a month. Three or four haystacks will get us through the winter without any shortage.

You have to be careful that the haystack is made of very dry hay. Too much residual moisture could cause it to ferment, which could lead to a fire. The fire could spread to the buildings. Our ancestors monitored haystack temperatures and did not hesitate to remove a stack if necessary.

Productivity of a Meadow

A meadow of 1 hectare can produce about 5 tonnes of hay. Half a hectare is enough to produce three or four stacks.

Grass produced by 1 hectare of meadow, fresh or in the form of hay, allows us to feed one cow or draught horse, over one year. Our Merens horse is satisfied with less.

▲ *Making hay is certainly one of the most beautiful collective worksites that one can offer to trainees or friends. Provided, however, that the area to be cut is not too large: we no longer have the strength and stamina of the farmers of yesteryear!*

▲ *Grain stacks were much more sophisticated than haystacks because they had to protect the precious food from rain. Threshing was carried out gradually over the winter. I would love to see such stacks, but unfortunately these skills have completely disappeared in France.*

▸ *These are the structures we make to support our haystacks. They are made of straight hazelnut tree branches and last for several years.*

The Quality of a Meadow

An old pasture can be of excellent quality. It may also have been compacted, acidified and have only poor vegetation. In this case, it may be desirable to plough it up and sow a mixture of fodder plants, mostly perennials.

In balanced grassland, there are many different plants: grasses (fescue, ryegrass,, timothy, cocksfoot, cattail, etc.), nitrogen fixers (clover, alfalfa, etc.), wild plants (yarrow, cowslip, etc.), wild plants (yarrow, plantain, chicory, etc.), whose deep roots can fetch water from the subsoil in times of drought, when grass and clover dry out because of their shallow roots.

Fertility must be moderated: an excess or lack of nitrogen causes a drop in biodiversity. Grazing helps to increase the diversity of plants present.

If mowing is done, it is best to avoid doing it year after year on the same plot, as this would cause some plants to disappear and impoverish the soil.

A meadow should be harrowed once a year. This can be done with a horse. In Normandy, a large flat metal ring, or an old cart wheel is dragged along to level out the molehills and soil. These operations take place in early spring. This makes mowing easier.

Thistles must be removed. They can appear explosively after tillage. The best way to control them is to deadhead them just before flowering. Like brambles and nettles, they do not like to be cut and at this stage they have no energy to grow back.

In environments with more than 500mm of water per year, the forest is the most suitable environment. Open grassland tends to become more acidic. This is one of the reasons why we surround our meadows with hedges and plant them in meadow orchards. Mowing is slowed down a bit by the rows of trees, so we mow by hand between them. Farming is all about compromise!

▾ *The amount of hay that can be stored in such a tightly packed stack is astonishing. Three or four stacks are enough to cover our needs for the winter. The rain runs off the stalks when they are well combed vertically. The haystack can last this way for several months, but it is better to protect it with a tarpaulin.*

Making the Most of Hay

A herbalist friend, Marc Grollimund, spent two hours in a pasture and came back with a list of more than fifty different plant species. Our organic hay, which is made without the use of any motorised machinery, is of high quality. It smells of all the fragrances of a meadow in June, even in the depths of winter. To our surprise, star chefs have used it for recipes such as hay bread and hay butter. The flavours of the grass are transferred to the dish. The price of a kilo of our organic hay is about five times higher than that of a kilo of vegetables... Don't hesitate to offer yours to restaurateurs!

Just like growing wheat, making hay has been an integral part of peasant life since time immemorial. Although market gardening takes up most of our time, we are happy to integrate these other activities into our farm. It is always stimulating to search and learn! The collective work that takes place during the summer months gives us precious moments of sharing and camaraderie that invigorate our energy.

◄ *Our attempts to process wheat without oil led us to build this barn and rediscover the tools and techniques of the past... A lot of work for a little bread! But this quest leads us to become more conscious of our environment and gives us undeniable aesthetic pleasure.*

9

GROWING WHEAT WITHOUT OIL

Making bread is a great experience, a personal commitment motivated by deep generosity and a desire to share. It is a way of being perfectly consistent, of giving the best of oneself.

Roland Feuillas[39]

"Sow your wheat in muddy soil and your rye in powdery soil."

Olivier de Serres, 1620[38]

Readers may be surprised that we are dealing with the subject of grains in a manual mainly dedicated to market gardening, which praises the merits of perennial plants and no tillage on every page. Especially since modern wheat and breads are unhealthy, sometimes causing gluten intolerance. We eat very little bread, apart from organic whole grain bread made at the farm.

So why should we be interested in grains, especially wheat? Because this fascinating plant has helped shape our civilisation. One could speak of a co-evolution between wheat and the societies of the Near East and Europe, just as rice has accompanied the peoples of the Far East, sorghum and millet in Africa, and maize in America.

The following pages ask the question: is it possible to grow high-quality wheat to feed yourself and your family without the use of fossil fuels? In the 21st century, can we do what our ancestors did for 10,000 years?

Our experiment at Bec Hellouin Farm was the first time we tried, and our results don't compare well to high-volume grain farms. Certainly, the 200kg of wheat produced in our small field of 1,500 square metres, if we had to calculate production cost, would be the most expensive in France. But they are a symbol of the autonomy to which we aspire and are joyful to do at every stage: the soil preparation and sowing, the growth and pruning, the harvesting, the milling of the flour, the baking of the bread, not to mention the eating it. We wouldn't trade a meal in the best gourmet restaurant for the taste of hot bread from the wood-fired oven. Pride, sharing, health, beauty, freedom: what food is as meaningful as a loaf of bread produced entirely on your farm?

▼ *Perrine makes a straw tie. This is done by choosing a few long stalks that are twisted before tying the sheaf of wheat. A simple and ancestral gesture, practised since the beginning of the Neolithic period.*

How much longer can we continue to cultivate industrial wheat that is sprayed with pesticides? It is destroying our most fertile lands for white bread that lacks flavour and nutrients, and requires additives.

In a post-oil world, the ability to produce wheat at home, without sophisticated equipment, will be valuable, just as it was in the past during troubled times and conflicts. Our attempts at this are in their infancy, but we want to share what we have learned so far.

Trials at Bec Hellouin Farm

When we set up our farm, we carried out two years of trial grain plots in the family garden. The birds ate everything!

Then we attempted to grow wheat, an approach that goes back to ancient times and that we will describe later (p.101). Again, the birds enjoyed it all to themselves.

For the past three years, we have been trying to rediscover the techniques of **wheat cultivated with animal traction** that were practised in France until the 1950s.

Bread School

We were guided in our research by Roland Feuillas who is both a miller and a baker. Roland created Les Maîtres de mon Moulin in Cucugnan, where they grow ancient varieties of organic wheat. His golden loaves have exceptional flavour, and are easy to digest, even for people who are gluten intolerant. Their bread is highly acclaimed by customers. Nutritionists who study its flour are very surprised by the exceptional mineral and nutrient content.

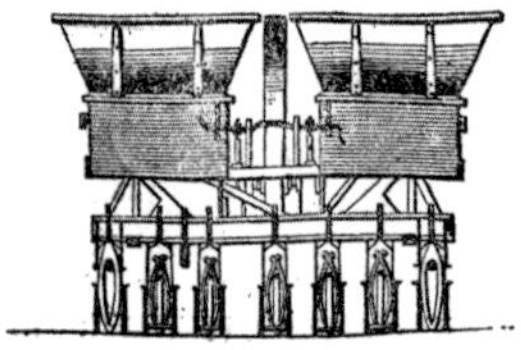

▲ *The 'august personality' of the sower, celebrated by the poet Victor Hugo, requires great mastery to obtain regular crop yields. The farmers of yesteryear designed machines to assist them. Our seeder at Bec Hellouin Farm is very similar to the machines shown in these engravings from 1835.*

Roland deems kneading bread by hand important on several levels:

Symbolic: in our civilisation, bread has been a symbol in many ways for over 10,000 years. An awareness of its power is a must for one's own level of humanity.

Social: daily bread fed the majority of our ancestors. Having this role with family and friends, it recreates and strengthens certain links. It activates and reinforces a connection to the fundamentals.

Spiritual: let us not ignore this dimension, let us not spoil this magnificent reality. Let's not be afraid to put it in its rightful place.[40]

Adapting Seeds to the Land

Roland gave us advice and also offered us seeds: a mixture of about thirty old varieties. At the same time, we received advice from Lydia and Claude Bourguignon. According to them, our soil is not suitable for wheat at all, but by adapting the seeds to our soil, we can increase the chance of having a crop.

In autumn of 2015, the mixture of these thirty old varieties, with high genetic diversity, was sown on a 1,500 square metre area with animal traction. Each year thereafter, part of the harvested wheat is put back into the soil. This process should result in a crop that is well adapted to the soil and climate of our farm. This should take about five years. The best adapted varieties thrive, while the less adapted ones regress. At the same time, they hybridise with each other at about 5% per year, to one day become 'Bec Hellouin wheat'.

▼ *Harvest time is a very important time for farming communities, often celebrated with festivals and religious ceremonies.*

In Search of the Right Tools

We have a lot of fun learning about the techniques and tools used by our ancestors. It's a form of ethnographic research. However, our approach is not a return to the past; we believe that by rediscovering pre-industrial knowledge, we will be better able to imagine post-industrial solutions.

We already had good animal traction tools for soil preparation and harrowing, as well as excellent scythes, and are well equipped with old tools for sowing, threshing and harvesting. Working with a horse is not necessary - you can produce wheat at home by hand on a small scale.

Soil Preparation and Sowing

In cold regions like Canada, wheat is usually sown in the spring, because it needs a warm summer to mature, but we grow autumn wheat, not spring wheat.

On the farm, soil preparation is done after the summer crop has been harvested or after a green manure crop. We avoid ploughing and prefer successive passes of the cultivator to loosen weeds and expose them to the drying effect of the sun.

Wheat prefers a slightly coarse rather than fine soil structure and care must be taken not to mix it too much. Wheat thrives in heavy, clayey, moist soils - yields are lower in light soils like ours. We fertilise the soil depending on its needs, with compost or green manure.

Sowing is done in September or October, possibly until mid-November. Early sowing allows the wheat to develop well before the cold winter and gives better yields, but it also encourages weed growth. Ideally, sowing is done at full moon.

It is possible to sow by hand. To achieve regularity it is preferable to stake the ground every 10 metres and calculate the quantity of grain to be sown for each square. Then it is advisable to make two overlapping passes. The amount of seed to sow by hand is about 200kg per hectare.

We have an old animal-drawn seeder. This beautiful equipment, in perfect working order despite several decades stored under a barn, allows multiple adjustments according to the size of the seed and the density sought. Our little horse pulls it without any problem. Seeding 1,500 square metres takes no more than an hour. With a mechanical seeder, the quantity of seed to be planted is around 140kg per hectare.

◄ *This old seeder is a little marvel of ingenuity. I am convinced that one day engineers will revisit these machines of the past to adapt them to a post-oil world. We would like to have a tool that allows us to sow with animal traction and avoid tilling the soil.*

After sowing, it is essential to harrow the soil, otherwise the birds will eat all the seeds. The depth of sowing should be between 2-5 cm. A roller should then be used if available.

Weeding and Tillering

When wheat is about 15cm high, a harrow can be used to weed. This momentarily disturbs the wheat but it recovers well.

In February or March, when young shoots are at the four-leaf stage, the plot can be grazed for a few hours by sheep. Grazing helps wheat grow new shoots, sheep fertilise and feed, and the crop will be better off. It can also be rolled or mowed.

Tillering is the stage when wheat grains can produce several stems, each bearing an ear. When a young shoot is planted deeper with ridging, nodes form with new roots and stems.

Current mechanised cultivation methods do not promote tillering. A single grain yields only a few stems, requiring a higher seeding rate. We will see below how garden wheat can produce a very large number of stems from a single grain.

Growth

Wheat doesn't need any special care for its growth. In organic farming we do not treat it at all. Hopefully, the weather will not be too wet and cause diseases, such as rust, to form. Wet weather also causes wheat to lodge, or lean over. This makes harvesting much more difficult.

Lodging can also be caused by strong winds and very fertile soil. Overly dense sowing makes the stems compete for light, thus weakening them.

Ancient varieties of wheat have fairly low grain yields. The stalks are taller than modern wheat. Both produce the same biomass per hectare but the ratio of grain to straw is different. Ours is great for bedding and animal feed. Rye straw is used to make thatched roofs, like the one on our house, and to make the straw roofs that cover market stalls. Tall stalks indicate that the grain was not hindered by weeds.

Modern wheat, with its heavy ears, has very short stalks, developed by selection and with the help of chemical stalk shorteners (yuck!). These short and uniform wheat stalks are much less prone to lodging but affected by weeds, so they are marketed with their own unique set of weed killers. They are also more prone to cryptogamic diseases, hence the use of pesticides.

Protection against Birds

When the ears of wheat ripen and are still tender and sweet, birds love them. Since our wheat crop is isolated from any other similar crop within a radius of several kilometres and is located very close to a forest, the experts warned us that the birds will eat everything. We even considered putting up a net and planting stakes for this purpose, but scarecrows turned out to work. Losses were limited.

"To get an idea of the fatigue that threshing causes the worker, watch him manoeuvre the flail: He lifts this instrument at least 37 times a minute and controls it when it falls. At 10 hours a day, he strikes 22,200 blows with a fairly heavy instrument. Only strong men can be employed in this kind of work. The slowness of the threshing process is another major disadvantage of this method."[41]

Maison Rustique du XIXe Siècle, 1835

▲ *A scarecrow keeps the birds away.*

▲ *With Rose, we test an antique sickle. Rose operates a crank to adjust the blade while I use a pitchfork to gather the stalks.*

▼ *Our ancient wheat strains are almost as tall as we are. Straw is a valuable resource for making animal bedding, and a source of organic matter for market gardens. Chickens feed on the grain missed by the thresher.*

Harvest

When wheat is ripe it turns yellow and dry. This happens around the beginning of August in Normandy. We harvest when the weather turns dry. It is a real celebration, the highlight of the year for farmers. We gather a few friends in addition to the team and mow our 1,500 square metre plot. It's a wonderful shared worksite, with all generations combined. Harvesting by hand is not a chore for us, quite the contrary!

In the past, harvesting was done with a sickle. Working by hand, a first pass cuts only the ears. Next, the straw is cut. This approach lengthens the harvest and prepares it for threshing.

Harvesting with a scythe is very efficient. This marvellous tool needs to be handled properly and perfectly sharpened, as described on p.116. Our ancestors installed a frame to their scythe. This was used to group the stalks together and lay them on their sides, ready to be tied into sheaves, with a single movement. A simple branch of hazel or willow bent and tied to the side of the scythe gives an idea of the effect produced by this frame.

Once the stems have been mown, they are tied into sheaves. Tying can be done with precut strings, the quickest solution, but not the most creative. We prefer to bind our sheaves the old-fashioned way, using a few long stalks of wheat. We group, then twist the stalks a little to make them more flexible, then we tie the sheaf on the ground compressing the bunch. An overhand knot is tied. It is very simple and sturdy. When the sheaves have been set they are arranged in piles resting against one another to dry out. If the weather is good, they can be left on the field for up to two weeks. This makes threshing easier.

◄ *Wooden rakes are well suited to harvesting, as is the 'devil', the large metal rake seen in action in the background. It is the origin of the expression 'pulling the devil by the tail': needy peasants could not afford to lose a single ear of corn.*

▲ *Threshing the scythe blade, in the field, on an anvil driven into the ground.*

◄ *Tying a sheaf with a straw tie. The knot is a simple overhand knot, and the end of the tie is tucked underneath to secure it. Sheaves are arranged in small stacks to complete the drying process.*

▸ *Flail threshing is a long and tiring job. We have looked at ways to make it easier, such as bundling the stalks in cloth.*

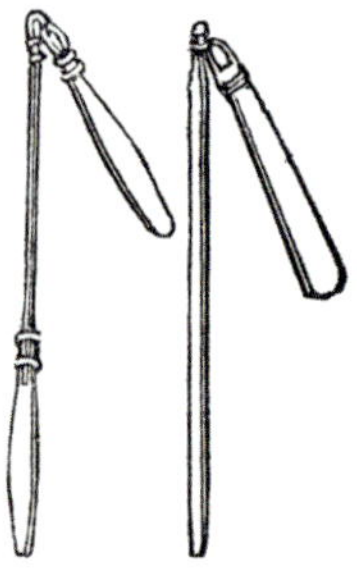

Flail Threshing

Separating grains from ears is the most difficult task. It is better to thresh as needed and let the grains ripen as long as possible. This way they will separate more easily. Our older neighbours describe flailing in the barn during pre-war winters.

Threshing is a simple procedure of beating stalks on a hard surface.

The flail was a widely used tool. Knowing how to use one is necessary because flails are not sold at garden centres. They are made with two sticks, a handle about 130cm long and 3cm in diameter, and a shorter, thicker hardwood stick to pound the stalk bundle. Two ash pickaxe handles will work. The two sticks are linked together with a leather tie. The link and its fastener must be very strong, because if the flail separates during the event, there is a risk of accident.

▾ *Threshing with animals is much faster than flail threshing. Mediterranean farms often had a dedicated stone area.*

To thresh with a flail, a threshing floor must be set up: a flat, clean and hard surface. We put a large solid cloth on the ground to collect the grain more easily. The sheaves are laid out in a bed about 20cm thick. The sheaves must be threshed for a long time, stirring them from time to time with a wooden pitchfork. This operation is tiring and very long. Doing it with several flails makes the work easier.

Here is how the authors of *Maison Rustique du 19ème Siècle* described flailing: "Several men can thresh together without harming each other, by arranging themselves two by two at a distance; they strike alternately, often in a rhythm, at the sheaves splayed out in front of them. The blows are made along the whole length of the sheaves, so that the tops of the shortest stalks are equally flailed. When one side is well threshed, someone turns the sheaves over, then, after having threshed this new side, he loosens the sheaves, forms a bed 4-6 inches thick, which they thresh and flip with the handle of the flail; finally, they shake the straw with the flail and threshes it again. Sheaves pass under the flail eight times."[42]

Threshing

We tried threshing by having our animals trample wheat, an operation once common in the south of France and Europe. We spread a large tarpaulin on the ground and turned our three horses around in a circle on it. It was much better than flailing; the grains were separated quickly and efficiently, with much less effort. According to the old-timers, another advantage of the threshing machine was that it crushed the straw better than flail threshing, making it more suitable for animals to eat.

Threshing with a Manual Machine

We then searched the internet, thinking that the farmers of yesterday must have developed faster solutions. We found a manual threshing machine that is still made today in China. It consists of a drum with metal spikes that rotate inside a housing with metal spikes, with only 1mm between spikes. When the drum is operated by a pedal or crank, the stalks are dehusked between the movable and the stationary teeth.

Just after we made this discovery, a friend gave us a machine of this type, made in the 1930s in Poland.

Winnowing

This is separating grain from the husk and dust left after threshing. The husks are the tiny skins that envelope the grains.

The simplest technique is to put grain in a large flat basket and throw them in the air on a windy day. The lighter chaff blows away and the heavier grains fall back in the basket. Light particles and straw, not blown away by the wind, will accumulate on top of the grain and can be removed by hand.

The **radial tarar** is an ingenious winnowing machine invented in the early 19th century. It has a fan in a box, operated by a crank. A fan blows away light particles, impurities and dust while the grains fall onto a series of mesh sieves that are shaken by the turning crank. The grains are sorted according to size. Antique winnowing machines are still frequently found in second-hand shops. It took some trial and error to understand how our two machines worked, but once we got the hang of it, winnowing became much easier.

Grain Grinding

Small mills with a stone wheel for home use can be found in shops or on the internet. Since flour loses most of its vitamins within a few days, it is best to grind it at the last moment.

The flour that comes out of a small mill of this type is integral, including the germ that is the part richest in vitamins, minerals and proteins. This makes high-quality flour. In industrial millings, sprouts that make flour nutritionally rich are removed, because they reduce shelf life. Bran can be difficult to digest – sift to extract it.

Baking Bread

Bread is made from a mixture of flour, water, salt and leaven. Leaven predigests gluten and removes phytic acid. It also improves the nutritional qualities of bread. Ancient wheat produces bread with exceptional nutrition, taste, density, and flavour. It can stay fresh for up to a week.

Young and old alike enjoy kneading dough. There are different ways of doing this, none necessarily better than the others, so look for the one that satisfies you. Such bread deserves to be kneaded by hand.

The bread can be baked in a domestic oven, or a small wood-fired oven. On the farm, one of the first structures built was an old-fashioned bread oven that can bake 50 loaves.

◀ *A friend gave us this beating machine. There is no one to show us how these old machines were used and we had to research it. The internet is a good ally for this type of ethnographic research, with videos available that have a wealth of information.*

For several years, we supplied a large Parisian CSA with vegetable baskets and bread. Twice a week I would do the cooking, at the end of the night, before the harvest. I loved to see the fire burning brightly while outside the sky turned pale as dawn approached. The whole orchard smelled of hot bread when it was baking. Our daughters were young and together we made bread in the shape of little animals. The bread is now baked by Fabien, the farm's chef, on the days when our shop is open.

It is all these steps that give the bread its magic, its strength. Like many things that can be done on a farm, making wheat and bread requires learning, time and effort, but the

▼ *Inside the tarar, or wheat sifter, there is a fan operated by a crank and a set of wire mesh sieves. This beautiful instrument is used to winnow and sort grain according to size.*

▸ *We built this bread oven in 2006. During the first few years, I made bread twice a week, to the delight of our daughters with whom we created animal-shaped buns that they found all warm and golden when they came out of the oven.*

▾ *Bread baked over a wood fire on the farm has an incomparable taste. Not to mention its smell: during the baking process, it spreads out across the orchard! We often suggested to our visitors that they take the bread out of the oven themselves, and observed how deeply this resonated with them and left a lasting memory. The industrialisation of our food has caused us to lose a sacred connection with the art of producing our own food.*

simple beauty of the sheaves, the gestures and the tools give life its true flavour.

Yields of Ancient Wheat

Ancient wheat varieties are much less productive than modern wheats. Yields are in the order of 15-20 quintals per hectare. When everything is done by hand, yields can be as low as 10 quintals. A lot of ears are left on the ground. In the past, the poor were allowed to collect them after the harvest.

Producing wheat by hand obviously excludes any form of commercialisation; you do it for yourself, for pleasure.

Being a Farmer-Baker

In recent years, a generation of self-taught enthusiasts has combined wheat production – generally old varieties lovingly exchanged from one farm to another – with milling and baking. These farmer-bakers produce exceptional breads, very different from industrial production. Their approach deserves to be supported.

The profession of farmer-baker is becoming more and more attractive. The farmland required for this type of activity is smaller than that required for traditional grain farming, and processing on the farm adds value to the crop. At the same time, there is a growing demand for organic and authentic breads, as consumers are looking for ethical products that combine respect for the environment with health and nutrition.

Although the activity is rewarding for the farmer-baker, it is nonetheless complex: in reality, they practise three trades, with many skills to acquire and investments to make. These investment costs are potentially high, but they can also be minimised and/or spread over several years, as illustrated in the table.

Many farmer-bakers build their own oven and knead by hand... In this case, the equipment needed is reduced accordingly.

Organic production allows wheat yields of around 30 quintals per hectare, and up to 80 quintals on some farms. These yields are lower with older varieties. For a quick estimate, taking as a basis a yield of 15 quintals per hectare of established wheat, three cultivated hectares produce 45 quintals of wheat. If we consider that 1kg of wheat allows us to produce 1kg of bread, it is then possible to produce 4,500kg of bread per year. Sold at a price of at least 5 euros per kilo, these loaves

ACTIVITY	AGRICULTURAL PRODUCTION	MILLING PLANT	BAKERY
Tools needed	Tractor or horses if animal traction is used Soil cultivation and fertilisation tools Sowing tools Weeding tools Harvesting tools Storage place for equipment	Sorting facility Grain storage area Milling unit (grinding mill or roller mill and sieve mill) Mill storage area	Kneading machine (mechanical or manual) Oven (electric, wood-fired, fixed, mobile...) Small equipment: scales, table, baskets, basins, layers, etc. Workplace (kneading and baking) Location for selling
Possible strategies to minimise investment costs	Contract work, subcontracting Pooling of equipment, exchange, cooperative for the use of agricultural equipment (CUMA)	Subcontracting Sharing	Renting an oven Sharing

generate a potential turnover of 22,500 euros per year. For sustainable production, these three hectares of wheat must be rotated every three years (nine hectares in total are then needed), ideally of five to seven years (15 and 21 hectares in total respectively). In order to practise this profession in a viable way on a microfarm of smaller surface area, one would have to achieve higher yields, better selling prices for bread (by enhancing its exceptional quality, by selling at the farm...), and/or diversify their activity.

A few years after setting up, the farmer-baker can obtain an average salary of 1,300 to 1,800 euros per month.

These figures are obviously given as an indication, as each project leader must size their activity and make goals according to objectives for income and labour. The profession of farmer-baker has an agricultural status, whereas the baker is a craftsman. The farmer-baker is therefore primarily a farmer who produces grains and processes all or part of them into flour and then bread. French regulations tolerate a threshold of 30% of flour coming from outside the farm. The job of a farmer-baker is a physical and skill-intensive one that requires a good deal of prior training and, ideally, internships with several professionals. It is possible to train in both cereal grain farming and baking.

For some years now, a certificate in farming and baking has been offered by the Centres de Formation Professionnelle et de Promotion Agricole (CFPPA) in Montmorot, in the Jura, and Coutances, in the Manche.[43]

Garden Wheat

We are now going to tackle a little-known subject that is nevertheless of great interest: the question of garden wheat. This practice dates back to ancient times.

How did the people who preceded us produce the cereal grains that were the basis of their diet? Our understanding of the past may lead us to believe that their archaic techniques were not very efficient and we might imagine that our ancestors chronically faced famine... It is hard to imagine that people could have survived before the invention of the tractor and the combine harvester!

How could the Egyptian, Greek and Roman civilisations have reached such a high level of refinement and erected monuments that we still admire without efficient agricultural techniques?

The cultivation of garden wheat requires only extremely simple hand tools, the same as those used for growing vegetables. It produced yields of up to 100 quintals per hectare, sometimes more – a level of production that today's cereal grain farmers are only able to achieve with huge amounts of inputs and fossil fuels!

A few years ago, we discovered the principles of garden wheat cultivation in the book *Agroecology* by Joseph Pousset.[44] We clumsily tested this approach, without giving it the regular follow-up it requires, and our grains ended up as bird food. Then we forgot about it for a while.

"In the present state of affairs in France, where threshing takes place almost everywhere without machinery, the threshing floor, where wheat and other grains are threshed, is an essential part of barn construction. It must be able to withstand the stamping of animals when the threshing takes place, as in southern Europe, and in this case, the threshing floors are almost always outside the dwellings; or to the percussion of the flail; in this case, the threshing floor is always placed in the centre of the barn."

Maison Rustique du XIXe Siècle, 1835[45]

► *Trial of garden wheat on a raised bed. The plants will be thinned to 30-40cm. Regular ridging and hoeing encourages tillering.*

In 2017, as we finished harvesting in the manner described above, someone passed me a small booklet written in French by Pierre Sauvageot and Paul Grillo, published in 1943, during the war: *La Culture Familiale du Blé*. It describes hundreds of attempts to revive this ancient practice with the aim of bringing health, food security and dignity to the war-torn people. This curious text mixes agricultural, hygiene and social considerations, but it is based on solid references. Here is a long extract:

"In ancient Egypt, two and three harvests a year were made of wheat 7 cubits high, of the species called *centigranum* by the Romans and counting 100 grains per ear (current average in France: 30 to 60 grains per ear).

"Several centuries before our era, China fed 450 million people and achieved wheat and rice yields of 80, 100 and 150 quintals per hectare (current average in France: 16.4 quintals from 1934 to 1938).

"Pline the Elder tells us that a proconsul from Africa sent the emperor Augustus a clump of 400 ears *from a single grain*. This represents a minimum yield of 12,000 to one. He also tells us that the Emperor Nero received, from his imperial estates in Byzacium, a tuft of wheat bearing 360 ears. In 1851, Major Hallet, Brighton, obtained 20 to 120 ears of wheat, each bearing 60 to 120 kernels, for each grain sown.

"In 1875 a single plant, bearing 105 ears and yielding a total of 8,000 grains, was exhibited at the Maidstone Farmers' Club, London.

"In France, well-known and reliable agronomists, such as Duhamel, Davy, Neufchâteau, Tessier, have seen wheat plants bearing 100 to 300 ears...

"The 175 kilos necessary for a man's annual diet were produced at the Tomblaine station on an area of 210 square metres [around 1880], which represents a yield of 80 quintals per hectare. But much more can be obtained from shatter transplanting [a technique that consists of transplanting the wheat plant several times, dividing it into shards - stems and roots - each time].

"In 1775, John Miller had already harvested from a single seed, a crop of 21,109 ears containing 576,840 very fine grains of superior quality, weighing 21 kilos 586 grams.

"This experiment of transplanting a single grain of wheat by separating stalks, repeated in 1903 by Mr Serrant-Bellenoux, yielded 709,700 grains, weighing 31 kilos 795 grams."[46]

And the text continues, mentioning experiments conducted by one of the authors that produced a yield of 160 quintals per hectare in a garden crop. Are there not similarities with the biointensive practice of market gardening? By intensively tending a small area, it is possible to achieve yields that seem inconceivable with conventional techniques.

The Secrets of Garden Wheat

How was it possible for farmers working with their bare hands or with a wooden plough to achieve such high yields? The cultivation of garden wheat uses a radically different method from the contemporary approach. Seeds are planted one by one, widely spaced and regularly ridged and weeded, which favours tillering and explains the number of ears that a single grain can produce. A single plant can reach a diameter of 30cm.

This method requires as much care as vegetable crops, but it can be practised in a small home garden. If an area of 210 square metres could produce the 175kg of wheat that fed a man in the 19th century for a year, nowadays, with the bread ration greatly reduced, a few dozen square metres could produce enough wheat for one person.

This ancient approach makes it possible for almost everyone to produce their own bread. A garden of a few hundred square metres would enable a family to become self-sufficient in both vegetables and cereal grains. This prospect is really fascinating and we will be doing it.

▲ *Tillering is when several stalks and ears grow from a single seed.*

Soil Preparation and Planting

It is worth noting that the authors of the booklet quoted were strong advocates of organic, and even biodynamic, farming and natural health.

They advocated no-tillage soil preparation, rather than successive passes of tine tools such as cultivators or ploughs, and fertilisation with compost or green manure. Furrows were made every 50cm or so. A grain of wheat was planted about 5cm deep every 30cm in the bottom of the furrow, not on top of the ridge as one might be tempted to do. Other seeds were planted in the nursery, to be transplanted to fill any gaps.

All the grain could also be sown in the nursery and transplanted, which also encourages strong tillering. Today, John Jeavons' raised bed method advocates transplanted grain.

The preferred seed was bearded durum wheat – the beards repel birds to reduce loss.

The wheat was sown as early as possible, from the end of August. Some farmers grew their crops without rotation, on the same plot, year after year.

Weeding and Ridging

As soon as the plants were strong enough, a first ridging was made using the soil left in reserve between each furrow. If possible, three or four ridges were made before winter. These operations were like hoeing to keep the soil clean while forcing each wheat plant to emit secondary stems, called tillering.

Hoeing and ridging continued in the spring for a total of five or six passes, accompanied by watering if necessary in dry weather.

The subsequent operations, harvesting, threshing and winnowing, were carried out as described in the first part of this chapter.

There is therefore nothing mysterious about the practice of garden wheat. It is based on the natural ability of the grain to grow and on intensive care. The savings in seed were significant, since the 200kg of seed per hectare for broadcast sowing were reduced to... 3kg per hectare! This saved seed alone could feed one person for a year, and compensated for the extra work.

As a permaculturist, I wonder how it would be possible to revive this very old practice, by combining it with current soil knowledge. Mulching between the rows? Replanting in the spring with another crop or a green manure?

I wanted to share this approach with you, even though we haven't really tested it yet – which is a bit contrary to the principle that we should only deal with subjects we know about in this manual – because I found it really interesting. In Germany, modern wheat cultivation trials are currently being carried out with sowing in rows 30cm apart, and yields are said to be higher than with tight sowing.

We should also mention the experiments of Marc Bonfils. In the north of France, he forced tillering by stumping wheat plants. His technique was based on:

1. Very early sowing, at the beginning of the summer preceding the harvest year.
2. A low sowing density in order to obtain one plant every 60cm.
3. A permanent cover crop of white clover.

The next wheat crop was sown into the previous wheat crop, which was still in place and maturing.

Marc Bonfils claimed to obtain yields of 150 quintals per hectare.[47]

In a post-oil world, it will be easy to produce vegetables and fruit, and even meat and dairy products, but what about cereal grain?

The garden wheat technique gives hope that we can continue to produce wheat, certainly not to feed livestock, but enough to provide good bread for local communities.

We will continue our experiments and would be grateful if you would share any knowledge you gain.

▲ *Growing wheat also makes people grow. It builds pride in one's ability to do things on one's own, confidence in one's abilities and a love of teamwork across generations.*

▼ *The sight of sheaves of wheat and sacks of grain awakens very old memories. We reconnect with our farming ancestors who anticipated the precious harvest on which their survival depended.*

AUBERGINE
POIVRON
DE
CHINE
COURGETTE
DIVERS
FEVE
HARICOT
JEUNES
POUSSES
POIREAU
POIS
RADIS
SALADE
TOMATE

XV

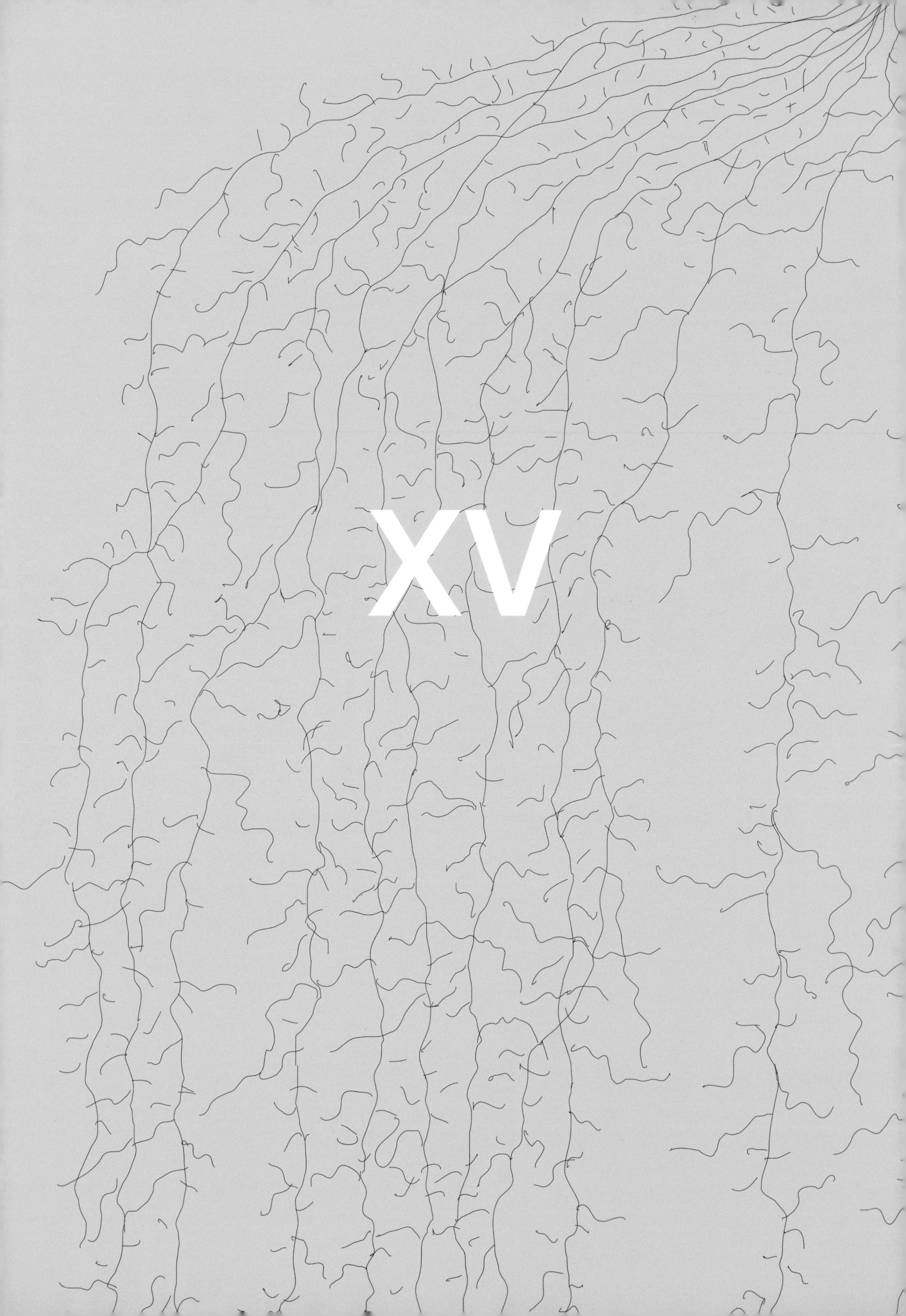

TOOLS

◄ *Blacksmithing is a difficult but extremely interesting art, there is something magical about seeing metal become flexible. We aim to learn the basics of it in order to make certain tools or metal elements for our projects.*

10

THE TOOLS

A convivial society is one that gives people the opportunity to act in the most autonomous and creative way, using tools that are less controllable by others... Productivity is expressed in terms of having, conviviality in terms of being.

Ivan Illich[48]

In ecoculture I find many elements that appealed to me when I was a sailor: learning to manage in all types of situations with the resources available, looking for the simplest solutions, which are generally the most effective... A farm is like a big boat in the middle of the land!

Whatever the farming system chosen, tools play a determining role in the efficiency of a farm. At Bec Hellouin Farm, we have made a radical ecological choice: to work the permanent cultivation beds entirely by hand and to use a draught animal for the storable vegetables that require a larger cultivated area. Thanks to the excellence of our manual tools and the coherence of the cultivation system, we manage to produce, in value, as many vegetables per hour worked as our colleagues with a tractor, with less investment and maintenance costs since our tools never break down.

► *These are the tools that have been my companions for many years. The aesthetic and sensory pleasure is there: you will find very little plastic on the farm.*

In Search of the Best Tools

Hand tools are an extension of the human body, and at the service of people and their projects, unlike many machines. We tend to think that progress liberates us by putting wonderful machines at our service, without realising the extent to which we have to work for them, sometimes to the point of becoming their slaves. The proof is that we have never been in such a hurry and under such stress as in the modern age. Those who are interested in this issue will benefit from reading Ivan Illich's *La Convivialité*.[49] He denounced the excessiveness of industrial machines that enslave human beings, whereas manual tools promote autonomy.

The question of manual tools has fascinated me since youth. I had the opportunity to work for several years in the workshop of a marine carpenter on the island of Noirmoutier, to build and restore boats. He was an extraordinary man who had a strong relationship with his tools. I sometimes caught him talking to them.

To carry out market gardening entirely by hand, in an era dominated by mechanisation, may seem incongruous. But when we first set out on this path I had great confidence in the efficiency of hand tools, inherited from my years on wooden boats. Long-distance sailing taught me that we needed the best possible tools, even if it meant creating them ourselves. Mediocre equipment on a boat is not practical because it will quickly break and you will have to buy a better one. It is better to get it right the first time. The best equipment is often the simplest and we are strong

▲ *Having a small anvil provides many services, for example, straightening a bent tool.*

supporters of low-tech. On *Fleur de Lampaul*, one of the last sailing schooners in Europe, on which I sailed for seventeen years, I observed that everything that was original – the hull, the engine, the rigging, which we had rebuilt identically – almost never posed a problem and could, if necessary, be repaired by ourselves at little cost. On the other hand, all the new, modern and sophisticated equipment, which was also very expensive, broke down regularly and took up 90% of the time dedicated to the maintenance of the boat.

There are many similarities between a farm and a boat. We have investigated the issue of hand tools at length and continue to do so. Our research is conducted along several lines.

The Tools that Existed before Mechanisation

The first line of research is to find out what farmers used to work with in the past. In the past, there was a culture of manual tools that has largely disappeared. People are more comfortable now with computers than working with their hands.

The ancients had few tools, but they were very strong because they were forged by hand, and much more durable than today's tools, which are usually made of pressed metal. They kept them all their lives, before passing them on to their children. Solidity and durability: this spirit is the opposite of today's tools, which are practically 'disposable tools', sold at the lowest possible price in huge specialist supermarkets.

Former farmers had a relationship with the local blacksmith that enabled them to order tools adapted to their needs. Each region built tools to its particularities: soil (rocky, heavy, light...), crops, etc.

▼ *Here are a few old gardening tools, hand-forged and indestructible, bought for a few euros at flea markets. The long, pointed hoe is handy for digging deep furrows for transplanting leeks. The small heart-shaped hoe works gently in rows of tightly packed vegetables...*

Farmers generally made their handles themselves; they knew how to replace them in any case. The handle is a very important part of a tool, and determines its ergonomics and efficiency. Everyone could have handles that were adapted to their body shape.

Our ancestors were too poor to afford to waste any metal. A tool was a precious possession. It was therefore maintained, sharpened and greased to protect it from rust. When you went to work, you had a truly reliable, handmade companion.

This spirit was taught to me by my friend, a marine carpenter, who himself worked with his father's tools. This traditional approach to tools prevails at Bec Hellouin Farm. We learn to forge, to make our handles. At flea markets, I am always on the lookout for old tools to study their shapes and sometimes put them back into use in our gardens to test them.

However, we are not locked into a backward-looking approach. Not all the tools of the past are suitable for today's market gardening. They are often unnecessarily heavy. Permanent bed cultivation requires specific tools that are generally smaller and lighter than those used by market gardeners of the past.

Contemporary Hand Tools

We are always on the lookout for contemporary creations and keep an eye on developments. Eliot Coleman, the great pioneer of organic vegetable gardening, has created many tools for small farms. Eliot generously gave us some wonderful tools: small hoes, mesclun (young salad) harvesters, a micro-cultivator to work the surface of the beds, and above all his famous 6-row seeder.

Japanese Hand Tools

Along the way, we discovered traditional Japanese tools that represent the ultimate in gardening. Never has the quest for excellence been taken so far as in Japan. A job well done is a founding value of that society. Tools are respected and have a sacred dimension.

The art of forging was brought to its peak by the craftsmen who once made the samurai's swords. When the carrying of swords was banned at the end of the 19th century, many blacksmiths converted to woodworking and making garden tools. Some of their descendants are still active. The special feature of the blades of many Japanese tools is that they are made using the Damascus technique, which involves combining layers of metal of

◂ *This small straw chopper is very handy for preparing nettles that are added to the feed ration of chickens and sheep to provide minerals.*

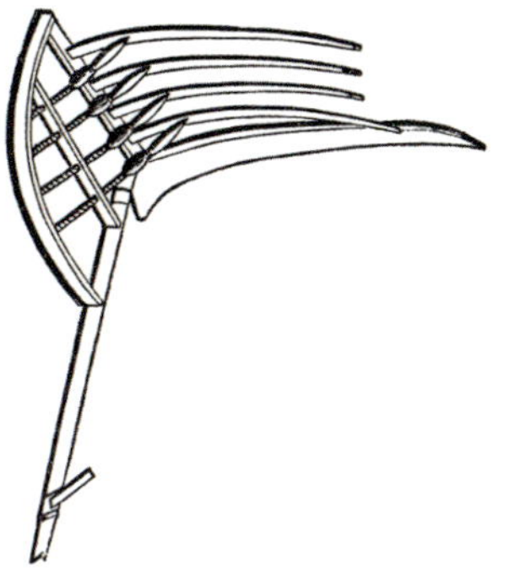

different hardness. The edge of the blade is made of an extremely hard steel, which gives it an incomparable sharpness but makes it brittle: the harder the steel, the better it can be sharpened. The central fold of hard steel is surrounded by folds of soft steel which give the blade flexibility. On sophisticated blades such as Japanese planes, the number of folds can reach several thousand!

One does not expect such finesse from a garden tool. Yet many Japanese sickles and harvesting knives are still made on the same principle. Once you have been lucky enough to use one, there is no turning back!

Japanese tools can be ordered on the internet. A nice range can be found in France at Coutellerie Champenoise. A few specialised sites in the United States offer a comprehensive range of Japanese tools for gardeners and woodworkers.

The Price of a Tool

A high-quality tool, handmade by a craftsman, is obviously more expensive than a metal tool stamped in a factory. However, the price difference is less than you might think. Junk tools also have a cost, and you have to change them often. It is easy to calculate whether it is better to buy a €30 pruning shear every year, which will do a mediocre job, or an excellent €110 Japanese pruning shear, which is remarkably ergonomic, powerful and sharp, and which you will keep for life. As soon as you go without motorised equipment, you can afford the best manual tools, which are incomparably cheaper than a combustion engine.

Moreover, you don't need many tools. For our research programmes I have bought a lot of hand tools for market gardening in order to compare them, but I still work with just a few simple chosen tools.

Tools Designed at Le Bec Hellouin

In most cases, the tools available on the market, especially Japanese and North American tools, meet our needs. But preparing permanent beds and establishing crops by sowing or transplanting remained problematic for us. We had to develop new solutions, so we created a bed preparation tool called the campagnole, some transplanting templates and a few planters of various types.

▾ *Cutting tools are of great importance on a farm. Farmers often added attachments to their scythes in order to group the sheaves together during harvesting. We test tools of various shapes, each with its own use.*

▸ *The campagnole is a wonderful tool, just the way we like them, as simple as it is effective! It is also the result of a friendly collaboration with a talented young craftsman, Vincent Legris.*

The Campagnole

The jewel in the crown of tools designed at Bec Hellouin Farm is certainly the campagnole! Its story deserves to be told. When we began our study of permaculture market gardening, we found that with a broadfork and a rake, it took us an average of 12 minutes to prepare 1 square metre of permanent flat bed (prepare in the sense of restoring; the initial creation takes longer). The broadfork is a wonderful tool for decompacting soil, but it has the disadvantage of raising clods that must then be broken up. This is time-consuming and tiring. These clods often fall into the pathways, so you have to lay down ropes and shovel and rake the soil back up onto the boards. Also, no broadfork is 80cm wide (the size of our flat beds), so two passes are needed to decompact an entire bed. Twelve minutes is far too long when you're alone on 1,000 square metres!

We looked for solutions. An 80cm wide rolling broadfork was made, but it was not practical to use. Then Perrine came across a video in Russian showing a priest, wearing a black cassock, who had the idea of adapting horizontal tines to a tool with vertical tines that looked like a small broadfork. We then contacted a young craftsman, Vincent Legris, and suggested that he work on the creation of a tool adapted to our permanent beds based on this idea. A beautiful collaboration between this craftsman, who is as kind as he is talented, and our team ensued. Over the next two years, Vincent produced a dozen successive prototypes that were tested in our gardens. The result is the campagnole, a wonderful tool that fully meets our expectations.

The principle of the campagnole is simple: the vertical tines loosen the soil without turning over the horizons, like a broadfork. We opted for a tine depth of 24cm, which is well suited to vegetable crops. A frame with horizontal tines is placed on the ground. The soil is lifted up by the vertical tines when the two handles are tilted backwards, and passes between the horizontal tines, which refines it. The horizontal frame levels the soil and shapes the bed into its correct size. The work is so precise that hardly any soil falls into the paths and a quick rake is enough to complete the seedbed.

Two wheels allow the campagnole to be moved along the bed without over exertion. Vincent added a pedal that allows the tool to be tipped over so that it can be easily transported between the workshop and the garden, like a wheelbarrow.

The speed is incomparable, even faster than a medium-sized tiller, and the campagnole is less tiring because the tiller tends to sink into the very soft soil of the beds. It takes about a minute to decompact 1 square metre, plus a final rake. Switching from the broadfork to the campagnole has saved us several weeks of work each year.

The campagnole comes in three widths. The largest is 80cm long, which means that flat beds can be prepared in a single pass. Vincent offers the campagnole with two types of horizontal tines, fixed or flexible. The latter are recommended for rocky soil (stones rise up and pass through the flexible tines,

▾ *Demonstration of the campagnole at the Shongaï Agroecological Centre in Benin. The campagnole is free of rights and can be reproduced; its plans are open source on our website. Widespread use of this tool would allow many small farmers to do without combustion engine tools.*

◄ *The largest campagnole model is 80cm wide, the size of our flat beds. It is so easy to use that these two photos are enough to describe it. The model shown has flexible, horizontal tines, well suited to rocky soil. A line along the side of the bed helps to guide the tool.*

▲ *The campagnole is easy to transport; it can be tilted and rolled.*

allowing the soil to be broken up afterwards).

This tool is cheaper than the smallest cultivator, is wear-free and works without a drop of oil and no maintenance costs. The campagnole is ideal for market gardeners who want to practise their profession in a way that respects the soil. It allows you to create an ecological and efficient farm with minimal investment.

The campagnole is marketed by Fabriculture, Vincent Legris' workshop.[50] The plans are open source. However, we suggest that you order your campagnole from Vincent rather than making it yourself. You will support this young craftsman and his effort to create innovative tools, and it will be cheaper. Indeed, many adjustments and technical details have been improved over the different generations of campagnole, such as the choice of steel alloys for its different parts, for example. Vincent's tool is very accomplished.

We are convinced that the campagnole is a leap forward comparable to the invention of the broadfork in its time. We find the two tools complementary: we still use the broadfork to decompact rounded beds.

Transplanting Templates

We transplant thousands of plants a year. Speed and accuracy of the transplanting is a major factor in the economic efficiency of the farm. Noticing how much the performance of the people involved in our gardens varies, we made transplanting templates, already discussed in volume two. They consist of a sheet of plywood into which wooden cones are inserted for digging the planting holes. These cones, or spikes, are simply made from hardwood tool handles. To ensure that they are firmly attached, they are driven a few millimetres into the plywood sheet, glued and screwed in place from above. They are made in the workshop so that they can be

▼ *Transplant templates*

positioned to the nearest millimetre, in staggered rows in order to increase plant density and consistent spacing. Two wooden handles are attached to the top of the plate.

Ideally, you should have a set of plates with pins spaced at 7, 9, 11, 13, 15 or 20cm depending on the spacing of the plants you wish to transplant.

A Few Tips on How to Make Them

Ideally, the trays should be half the width of a flat bed.

The staggered arrangement of the pins should be made in the shape of an arrow, so that the successive positions of the plate follow one another without leaving any missing holes.

To reposition the plate quickly and accurately after each application, the distance between the edge of the plywood and the edge of the first pin should be the same as the distance between the pins, so that by aligning the edge of the plate with the last marked holes, the following holes are positioned at the right distance.

Do not make plates with more than 12 pins, as they would be too heavy and difficult to press in.

How to Use

The template is placed on the ground and pressed down – stepped on if necessary – to drive all the pins to the desired depth. Then the handles are grasped and a few small up and down movements are made to mark the holes before the plate is removed.

Soil moisture content is important, just as it is for the multi-row seeder:

- If the soil is too dry, the edges of the holes will crumble. In this case, pour one or two watering cans over the bed and wait about 15 minutes for the water to penetrate.
- If the soil is really wet, the soil will stick to the template... Better to wait until it dries!

We carry a hand planter, or dibber in our pockets in case we need to re-form one of the holes.

The seedlings, in mini clumps, are simply pushed into the holes with a quick movement. With the templates, it is possible to transplant about 400 plants per hour.

These templates can also be used to simply mark the holes where the seeds will be placed for direct sowing. In this way the seeds are very precisely located.

▾ *Dibbers*

Dibbers

Dibbers are wooden planters of varying diameter and length, with a hand guard that resembles that of old swords. This guard, or stopper, acts as a depth gauge when the planter is pushed in, while at the same time digging a small trough in the soil. When transplanting on the side of a bed or at the top of a ridge made with animal traction, a trough is useful for catching water that tends to run off the slope and increases the efficiency of the first few waterings.

T-Planters

Jean-Claude, who has been working on the farm for more than ten years, has made a series of wooden dibbers from a 50cm-long piece of round wood (a tool handle) with a T-handle. The long part is graduated every 10cm. This type of planter is practical because it allows the spacing between plants to be determined instantly. Its length makes it possible to dig holes while standing.

A Complete Range of Tools

The campagnole is best combined with the other hand tools designed for flat beds, mentioned in other chapters: an 80cm rake, the 6-row seeder, micro-hoes, manual

harrow-tillers... This provides a very coherent range of tools. This collection could be enriched with a multi-purpose tool holder, with a wheel spacing that allows it to pass over flat beds. Such a tool would not be very complicated to build but we have not yet found the time for it. It would be useful for transporting crops, spreading compost or as a frame for attaching light tillage tools.

I will introduce you to some contemporary tools that are essential for our farm.

The Multi-Row Seeder

The 6-row drill-driven seeder designed by Eliot Coleman is very efficient: when he used it as a replacement for earlier versions of professional seeders, Eliot saw his yields almost double.

We use it mainly for small seeds, especially carrots, turnips and radishes, because in our growing system we transplant all the vegetables that can be transplanted.

Setting up the seeder is not easy. As already mentioned in volume two, before any sowing in the field, we recommend using a nice long table, covered with a light white cloth, on which you can roll the sowing machine to see the flow of the seeds. Start by adjusting one bucket and then adjust the other five when the seed flow from the first bucket is satisfactory. Make a note of your settings for each seed type to save time later.

We are currently working on improving Eliot's seeder, with his blessing, as there are some areas that can be improved.

Micro-Hoes

We have a nice range of hoes in different sizes, including very small ones, which are very useful for going between tight rows.

Manual Harrow

Johnny's Selected Seeds sells a manual weeder that is used to destroy weed seedlings after the crop is well established. This saves a lot of weeding time.

Mesclun Harvesters

The same American distributor offers very practical mesclun harvesters. Provided that the blade is kept sharp and that sowing is dense, the multi-row seeder is perfectly suited.

▲ *From left to right, tools adapted to flat beds: precision hoes, 4-row seeder, 6-row seeder, wide rakes.*

The use of the tools presented in this chapter is detailed in volume two of this manual.

▸ *To use a scythe efficiently, the tool must be adjusted to size. When standing with arms relaxed, the heel of the blade should rest on the ground, its tip lifted off the ground by about 1cm, as should the cutting edge. You can also adjust the opening of the blade in relation to the handle, so as to cut more or less grass with each stroke. Beginners are well advised to close this angle.*

The Scythe

We will now describe in detail a fantastic tool that has unjustly fallen into disuse since the invention of motorised brushcutters. Anyone can use a brushcutter, whereas the scythe requires a real apprenticeship, especially for sharpening. The truth is that handling a scythe is an art. Once mastered, this environmentally friendly tool is very satisfying – it can even be addictive! A well-handled scythe allows you to work as fast as with a motorised brushcutter, in silence, enjoying the scent of the cut vegetation, without incurring any expense and without damaging the planet! The scythe is invaluable for gardeners and farmers, for destroying green manure or old crops, for mowing mulch or fodder for animals, or for farm maintenance.

There are two types of blades that can be mounted on the same handle.

The sickle is a relatively short (less than 55cm), strong blade designed to cut all thick-stemmed vegetation, such as nettles, brambles and ferns. The blade is stone sharpened, not beaten like a scythe blade, so that the cutting edge resists woody stems or the impact of stones. The mower is a powerful tool, well suited to the maintenance of a forest garden, embankments or ditches.

The scythe itself is a long, very thin blade, designed to cut the fine stems of grass and cereals. It is also very effective for green manure and old crops. A scythe blade is sharpened by beating it, as described overleaf, which gives it an extremely fine cutting edge.

Scythe blades come in different lengths, the longer ones being difficult to handle and reserved for very experienced mowers. For beginners, a 65cm blade is ideal.

The Best Scythes

Older scythes were often heavy and difficult to adjust. Modern scythes found in garden centres are almost impossible to sharpen properly because of their thick blades – manufacturers assume that today's users can't beat, so they offer an indestructible but inefficient tool!

Fortunately, it is possible to find good scythes, probably much better than the old ones. In Austria, and generally in mountainous countries, the small size and steepness of the plots have kept the use of scythes alive long after they have disappeared from other agricultural regions. The Austrian company Flux still makes little marvels today. The blades are forged by hand. The handles are made of bent ash and are light and strong, while offering multiple adjustment options. These tools are imported into France by our friend Emmanuel Oblin, a passionate person who looks for the best equipment, which he offers on his website (www.comptoirdelafaux.com). There are also tips and demonstration videos. The American site Scytheconnection (www.scytheconnection.com) offers many resources in English, and Austrian scythes in the UK can be bought from Permaculture Market (https://shop.permaculture.co.uk/collections/tools).

Austrian scythes are also favoured by English-speaking specialists, where this fine tool is making a comeback. The cost of such a

▾ *Top: Sickle blade, short and thick. Below: Scythe blade, long and very thin.*

◄ *The blade must remain in contact with the ground throughout the movement (unless there are stones, molehills...); a scythe should not be confused with a golf club! The legs are moderately spread, the knees supple, the back almost straight. The whole upper body accompanies the movement in order to gain power. After each stroke, the feet move forward only a few centimetres.*

scythe is much lower than that of a motorised brushcutter. It is a wear-free tool that you can pass on to your children.

Adjusting your Scythe

To work without fatigue or back pain, a scythe must be properly adjusted to size. The aim is to work with a straight back. To do this, while standing, start by positioning the intermediate handle at the height of the top of your femur, then adjust the upper handle to one cubit above (the distance between the elbow and the fingertips). You can then adjust it according to your physique.

The angle of the blade with the ground is adjusted while standing, by placing the scythe on the ground. The tip of the blade should be raised about 1cm from the end fixed to the handle. This adjustment is made by calibrating the height of the intermediate handle or by inserting a small wooden wedge between the handle and the scythe blade, in the metal collar that holds them together. The angle of the blade and the handle is also important. If this angle is open, the blade mows a larger area of grass and the effort required is greater. It is advisable to reduce this angle at the beginning. This can be done by adjusting the heel of the blade into the mounting collar.

Mowing Correctly

Good movement is essential if you want to mow without burning out in a few minutes.

To mow properly, the body should be upright, leaning slightly forward, with legs wide apart and knees flexible. The whole upper body is worked in a circular arc. The rotation of the whole torso and arms gives power to the movement. The right hand pushes down on its handle, while the left hand pulls. Weight shifts from one foot to the other.

The scythe remains on the ground, even if you want to mow low, during the whole movement. It is important not to make a golf movement! The blade glides over the grass along its entire length, which allows for an efficient and clean cut. This is achieved by mowing only a small portion at a time, no more than 10cm deep, but over a large width – the entire width of the arc of a circle that starts to the right of the mower and ends to its left. The advance is made by slightly moving one foot after the other a few centimetres, without taking them off the ground.

Depending on how comfortable you are, you can mow in a larger or smaller arc. Start with one metre wide, then move up to two metres once you are well trained. The width of the arc also depends on the density of the vegetation to be cut.

Peening a Scythe

Sharpening is done by peening followed by stone grinding.

Peening the cutting edge with a hammer stretches the metal. This task requires

▸ *A scythe is peened, otherwise its sharpness remains mediocre. Modern anvils make this operation easier but learning this particular skill requires training.*

▴ *Peening hammer with an antique anvil, a beautiful object that was driven into the ground at the workplace.*

exceptional finesse. We use a special peening hammer and a small anvil. All sharpening supplies are available online.

Our early ancestors used anvils that were staked into the ground. This made it possible to sharpen a scythe in the field. We tried it, but using this type of anvil requires some practice. Modern anvils have guides to prevent the blade from slipping. We set the anvil on a block.

The movement for peening is difficult to describe - we suggest watching demonstration videos. You have to strike the cutting edge of the blade uniformly in a way that stretches the metal. Progress slowly along the entire length of the cutting edge. A second pass is usually necessary. When peened, the blade is so thin that a fingernail can change the shape.

Striking involves a simple movement, where precision and uniformity are the challenge. Even sloppy peening, however, can improve a blade. Peening takes about twenty minutes, including disassembly and putting it back together.

Remove blade from handle to sharpen. Some people peen with the curvature towards the bottom, others do the opposite; the direction of the blade is not important.

Sharpening with a Stone

A scythe is usually peened about once a day. Between peenings, stone sharpening is done regularly, every five to ten minutes, with a wet stone. This keeps the blade razor-sharp. To sharpen with a stone, stand up straight, with the end of the handle set on the ground, and the blade at face level. The stone's shape is elongated, coming to a blunt point. It is used on the edge, not on the flat. The stone is placed against both edges of the blade, on the inside of the curvature. The support of the blade's non-cutting edge gives the correct angle for sharpening. Pull the stone downwards, diagonally, and repeat the gesture a number of times, progressing towards the tip of the blade. This operation stretches the metal and forms a slight fold of steel on the other side of the blade. Make a few passes on the other side to wear it down to the cutting point. The stone must be kept wet, and

▸ *Disassemble and clean the blade before setting it on the guide. Strike with the rounded end of the peening hammer, with short regular strokes along the cutting edge. Two or even three passes are generally necessary.*

◂ *To safely sharpen with a whetstone, keep your elbow in contact with your body. This will prevent fingers from coming in contact with the blade.*

▴ *A synthetic whetstone of coarse grain, for a machete. Fine natural stone for a scythe. An old horn whetstone holder, worn on the belt.*

free from molecules of steel that can clog its pores. The clogs make it less abrasive. The whetstone is worn on a belt, in a case filled with water. Our ancestors had very beautiful stone holders. They were made of cow horn or carved wood. Today we find them in brass or plastic. For a scythe blade, use a very fine stone. A stone that is too abrasive removes too much metal and makes the need to peen more frequent. A machete blade is thicker and requires a coarser grained stone.

Safety when Sharpening and the Use of a Scythe

The scythe is so sharp that any mistake can cause a deep cut. Always keep focused control of your movements. Be aware of your surroundings when working near others.

Stone sharpening is dangerous if done incorrectly. It is absolutely necessary to prevent the hand from moving parallel to the blade because a false movement could cause it to come in contact with the cutting edge.

To do this, ensure that the elbow of the arm with the whetstone remains in contact with your body at all times. This forces the forearm to follow an arc, and move away from the blade.

When working with a team, take care not to leave sharp tools on the ground. A running child or an inattentive adult could get hurt badly.

◂ *Position of the stone on the blade: maintain the correct angle by remaining in contact with both surfaces. Pay attention to your fingers to avoid accidents.*

Sheaths

This naturally leads us to talk about cases. One of our interns, Sébastien Fauterre, made superb leather sheaths for our scythes. These protect both the tools and us. A beautiful leather case is pleasing and puts tools conveniently on the belt. We rarely find cases well suited to specific tools, especially rare tools like harvest knives and pruning shears. I learned the basic skills of leatherwork and find making my own cases to be easy and deeply satisfying.

▾ *A very nice case, for an object that we enjoy maintaining and keeping forever!*

The 'Wheelbarrow-Workshop'

When working in the garden we take a few tools with us, then realise that we need something else so we return to the tool shed. Then we need something else and so on. We end up losing a lot of time commuting. Twenty minutes of unnecessary journeys per day waste almost two hours per week and two weeks per year. To limit the journeys, I made a 'wheelbarrow-workshop': simply a wheelbarrow on which I can secure my toolbox and long-handled tools.

▲ *This old wheelbarrow has been with me for 30 years – with a few repairs over the years. Metal wheelbarrows only last a few years and are difficult to repair. I added a small wooden structure that supports long-handled tools, and a place for a toolbox.*

Tool Bag

I use an old railroad worker's bag that allows me to carry just about every tool needed for an ordinary work day, apart from long-handled tools. With its different compartments, it is more practical than my old tool box, especially when it rains.

◄◄ *(Opposite page) The art of leatherwork and saddlery has not changed much over the centuries. The hand tools presented here should have elaborate cases. Leatherwork uses cutting, drilling, and sewing tools... Making leather cases for your tools is easy to do.*

◄ *This is my tool bag. Comfortable to carry, it really serves me well.*

Here is what the tool bag generally contains:

- Four different sized dibbers
- Three transplanting trowels of different sizes
- Japanese pruner with short blades
- Japanese pruner with long blades
- Two harvesting knives
- Japanese billhook
- Hatchet
- Japanese saw
- Pair of thick gloves
- Tape measure
- Labels
- Notebook
- Markers
- Lighter
- Excellent fixed blade knife and sheath
- Whetstone holder (horn case) and its sharpening stone
- Cloth
- Water bottle.

◄ *We made this one-piece donkey bench from a pine trunk. Its shape is fun, but a bench made with two planks would have been lighter and probably more ergonomic.*

A Few Useful Tools from the Past

I would like to introduce you to some tools from yesteryear, almost forgotten, that are particularly valuable on a farm.

Drawknife

Drawknives have metal blades that come in variable sizes and shapes, with their ends encased in wooden handles. Well sharpened, they have many uses, particularly for making tool handles. Pull it toward you, with the handles pointing towards the ground. The blade can cut long chips, but also do fine and precise work.

▲ *Drawknives can still be found at second-hand dealers, or they can be purchased new. It would be difficult to do without this tool. Use different size drawknives for various jobs: the big ones are powerful, the smaller ones are more precise.*

The Planing Bench or Shaving Horse

In the past, at farms and in craftsmen's workshops, a shaving horse was commonly found. They had a sturdy wooden seat and a foot pedal to push down and secure the work with hands free to work with the entire upper body, for great strength. Just release the pressure of the feet to move the piece. A long tool handle can thus be worked at any length with great comfort. On the farm, we made a donkey bench out of a single pine trunk. It's heavy. At the time, we found little documentation on this subject. Bernard Bertrand recently published an excellent book on shaving horses. He gives all the instructions needed to make one.[51]

Froe

The froe is used to split green wood. On farms of the past, this tool was often used for the manufacture of everyday objects. It has a metal blade fixed at right angles, set in a wooden handle. The blade is placed on the end of a branch or log, and driven in with a mallet; the handle is then moved from side to side to open the slot. Our ancestors worked a lot

◄ *Making shingles in the workshop. There is a strong link between microfarms and craftsmanship. Let's leave industrially manufactured products to big agriculture and instead turn to local artisans for our needs, or learn to do these things ourselves.*

with green wood – it splits much more easily than dry wood. Chestnut and hazel were often used for fences and baskets. Shingles from chestnut or oak were common in traditional construction.

On our farm, these three tools are frequently used. We make a lot of shingles to cover our small buildings. I also make crop labels for our plantings. I'm sure that with the decline of industrial civilisation, people will be interested in this ancient knowledge, as we can already see happening in England.

Tool Handles

The froe, the drawknife and the donkey bench allow you to make beautiful tool handles. Tool handles adapted to a person's size are ideal.

In Europe, the most common wood for tool handles is ash (*Fraxinus excelsior*). Ash is both flexible and resistant. For tools that are not subject to intense strain we can use a lighter wood, like linden or beech. A good handle should be as light as possible to minimise exertion.

To make a handle use the heart of the wood and remove the sapwood. The froe is practical for splitting a thick log into sections with each one destined to become a handle. Wood split with a froe is stronger than sawn wood because the splits follow the grain instead of slicing through it. Using the plane and donkey bench, work the piece of wood to give

▲ Splitting a chestnut log with a froe. For striking a metal tool, it is better to use a wooden mallet to avoid metal splinters. Wooden mallets wear out quickly, but they are easily replaced because they are made with tree branches found on the farm. The froe and the drawknife are all that is needed to make shingles that cover some of our buildings (this photo shows the barn roof).

it its shape. The finished work is done with sandpaper. Handles generally being worked as greenwood should be left to dry several months before fitting because drying reduces their size and creates a gap between the handle and the tool. It is helpful to have a small stock of handles ready to go.

The handle is fixed in the head of the tool. To ensure a tight and safe fit, the end of the handle is split and a hardwood wedge is jammed into the gap.

To make a curved handle, it is better to avoid sawing for the reason mentioned above. Start from a curved branch, or make a bending template that will force the handle to adopt the desired shape during the drying phase. The handle is left two to three months in the template.

Bending

Bending is achieved by boiling wood, or exposing it to steam, for a lengthy period of time depending on the diameter of the piece (approximately one hour per centimetre of thickness). Thus heated, the wood acquires great flexibility, bending easily while it is hot. We can make a fence gate or a basket handle by inserting a tool handle into a bending template.

It is easy to make a bending iron: solder a plate to the end of a big section of steel pipe. The pipe is inclined, and partially filled with water; light a fire underneath, and stop up the other end with rags.

Discovering this knowledge brings a lot of satisfaction and allows the farmer to benefit by making many objects that are as beautiful as they are useful, and all made from farm resources. Blacksmithing and basketry are other useful skills to learn.

Whatever the size of your garden, the right tools will accompany you for a long time, and be perfectly adapted to your body shape and your needs. This may lead you to modify some tools and handles to work more efficiently, and with less effort.

The appendix, p.260, offers you a detailed list of all the tools that we use on the farm, with cost given.

▲ *The clogger's knife is a long blade with a hook that attaches to the work bench. It is a powerful tool for rough cutting wood, the finish being preferably carried out with a drawknife.*

◂ *Steam softens wood with the release of heat and vapour.*

11

TOOL MAINTENANCE AND SHARPENING

It is sometimes said that the tool is the soul of the Japanese carpenter, just as the sword was the soul of the samurai. This comparison is accurate in many ways. To be more precise, the tool is the means by which the carpenter will leave a trace of his spirit in the soul of the wood, the link which will allow a marriage between their two souls.

Azby Brown[52]

Sharpening is a forgotten art, essential for manual tools.

◂ *Natural Japanese stones are little wonders, destined to last for generations. These pruning shears, hand-forged in Japan, are completely dismantled for sharpening. The stones have a decreasing level of grit, down to 8,000 for a mirror polish.*

Proper Use of Hand Tools

Using hand tools requires training. These tools are effective only when they are well chosen, used correctly and properly maintained.

In Japan, a country where hand tool culture is still strong, their usage is taught at school. Training for craftsmen is very demanding. A young craftsman masters his art when he knows the work to perfection and can move quickly. The extraordinary quality of Japanese tools is accompanied by a unique expertise in the art of sharpening. The best Japanese planes can make shavings several metres long, at a thickness of less than 10 microns. They are almost transparent. Knowing that the diameter of a hair is of 100 microns, this shows the potential of a hand-forged blade.

The best tool in the world is not worth much if you do not know how to use it, maintain it and sharpen it. Since this essential knowledge is no longer taught, it is up to us to find out how to do it. Market gardeners aspiring to have a strong ecological practice can make an exciting journey into the discovery of tools! This adventure is a source of great satisfaction. Most of us don't know that there is any difference at all between working with rusty, unsharpened tools, and perfectly maintained tools, specifically chosen as work companions. Of course, maintenance and sharpening require some time but this is compensated through increased efficiency on the job.

Level of Care Varies Depending on the Tool

Not all tools require the same attention. They can be classified into two categories:

- Tools intended for working the soil: spades, shovels, rakes, picks... They do not require much maintenance beyond a brush at the end of the day to remove dirt. If the edge is twisted from impact, it can be straightened by hitting it with a hammer on an anvil or using a grinder.
- Sharp tools, such as pruning shears, harvesting knives, pruning hooks and scythes, axes, and precision hoes benefit from being sharp. They require more care at the end of each work session and even while working.

The following notes mainly concern this second category of tools.

Metal Tool Maintenance

Almost all manual tools used by market gardeners have metal parts. In abrasive and humid environments like the garden, metal is subjected to harsh conditions. Environmental aggression has two consequences: corrosion and erosion of blade edges.

Corrosion Prevention

The first level of care is to prevent corrosion. Care is different depending on whether a tool

▲ *In the greenhouse workshop, an old grindstone restores the cutting edge of garden tools.*

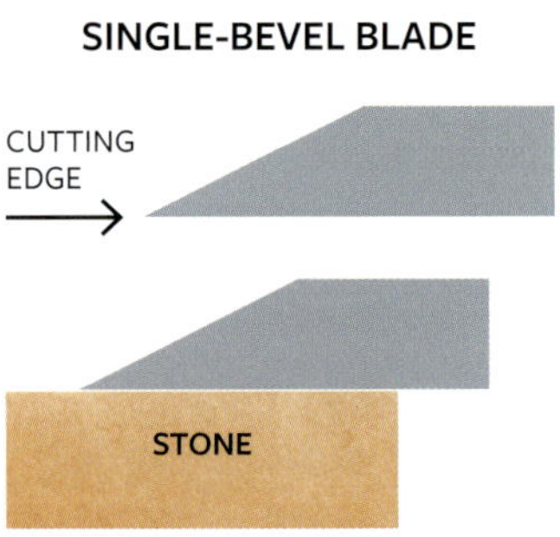

▲ *A single-bevel blade should be kept perfectly flat on the abrasive surface to avoid creating a second bevel.*

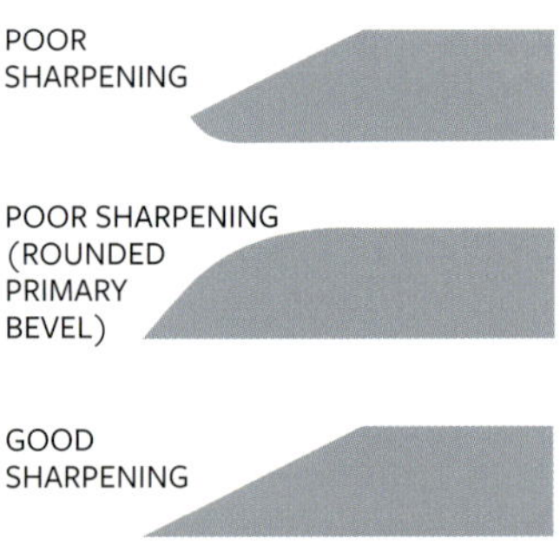

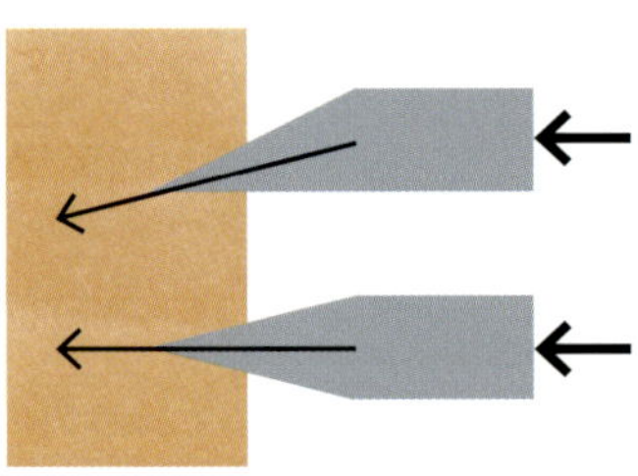

▲ *The tool path in the material follows the angle of the primary bevel.*

is made of steel, which rusts, or stainless steel, which will not rust if it is of good quality.

Stainless steel does not require special care beyond regular cleaning to avoid clogging the pores – plant juices make pruning blades dirty, for example. Since it won't rust, stainless steel offers a maintenance advantage. This advantage is mitigated by the fact that it is more difficult to sharpen. It also costs more. I prefer to use tools made with normal steel.

How fast rust happens depends on the carbon content of the steel. Contemporary tools are protected by an anti-corrosion coating that wears out over time, leaving chemical molecules in the soil. My preference is towards raw metal tools that require a little more care.

Tools should be cleaned carefully at the end of each work session. If we want to keep them free from corrosion, a regular application of grease or oil to protect the metal is needed. You can use a grease bomb or commercial oil, but these products are chemicals – the blades of almost all tools' sharp edges come into contact with plants that we are going to consume, in particular the secateurs and harvesting knives. The ideal is therefore to use edible cooking oil.

Personally, I have a stainless metal pot into which I regularly pour a little oil – not too much because it deteriorates quickly. A brush remains permanently in this pot. At the end of each session, I pass a quick brush stroke on my blades, then wipe them with an oily cloth. These simple gestures are enough to soak the steel with grease and effectively protect the blades.

At the end of the season, a thorough cleaning and a generous application of oil allows tools to winter under the best conditions.

One advantage of making the cases out of leather is that they become oily inside, as the leather retains oil, so the blades are slipped into an oily leather environment after each use.

Reconditioning a Rusty Tool

First remove rust with a wire brush and sandpaper. A wire brush mounted on a lathe and a belt sander attached upside down in a vice so that the abrasive strip faces upwards are invaluable aids. There are power tools available with a belt sander and a wire brush combination that are very useful.

Using the wire brush dulls the tool. You must then sharpen the edge.

Once the metal has regained its polish and edge, apply oil.

Sharpening a Tool

Sharpening a blade means removing metal by rubbing it on an abrasive surface, to restore its edge.

Learning to sharpen involves some basic knowledge – it applies to all types of tools and knives, and will be useful both in the garden, workshop and kitchen.

One or Two Bevels?

Some blades only have one bevel, and the other side is flat. This is the case for shears, wood chisels, drawknives and most planes. Other blades, like those of most knives, pruning hooks, and axes, have two bevels.

Identify whether you are dealing with a blade with one or two bevels. In the first case, it is absolutely necessary to avoid creating a bevel on the flat side. When sharpening, be careful to place the flat side of the blade perfectly flat on the abrasive surface.

Bevel Angle

The angle of the primary bevel depends on the intended use.

An acute angle gives a better cut but the edge is more fragile. The very thin blades, like those of harvesting knives, are reserved for delicate cuts of soft material.

A more open angle slices less but is more resistant. The edge of an axe, for example, or of a pruner, is stronger than that of a billhook.

When sharpening, it is important to maintain the original angle set by the manufacturer – except if you decide to change the behaviour of the tool.

Burr

Burrs are the thin film of metal created by sharpening the cutting edge. When we work one side of the blade, the bit tends to fold onto the other side. We remove it by working it gently with a fine stone, one way then the other. A pass over sharpening leather also eliminates it.

Sharpening Equipment

Since sharpening means removing metal, any abrasive material can be used – even a simple stone or a concrete step! But today there are many options. Materials used for sharpening are mainly:

- Stones, natural or synthetic (water stones, oil stones, millstones).
- Abrasive bands or discs, made up of a surface coated with various abrasive materials. The level of abrasiveness is described by a number which indicates the number of grains carried per unit of surface. The principle is the same for stones, grinding wheels and abrasive papers.
- The larger the grains, the greater the abrasive effect will be and the fewer grains there will be per surface unit. The number describing the degree of abrasiveness will therefore be low: 40 grit paper describes a powerful abrasive, useful for quickly making a first roughing. Papers 80-100 are medium grain and versatile. The papers with a grain ranging from 120-240 allow you to obtain a fairly good level of finish.
- The smaller and more numerous the grains, the finer the polish. Stones and small grained papers are used for finishes. Stones and abrasive papers rarely go beyond a grit of 400. If you want to sharpen to perfection, we can use stones a lot finer, at grain 1,000-3,000 and even up to 8,000, like certain Japanese stones. The blade is then polished like a mirror. For garden tools, there's no need to go that far!

When sharpening, it is best to use stones and belts of different grits. If the work is significant, start with a large grain to re-establish a damaged or rusty metal surface. No point trying to repair a very damaged edge with a fine grain, unless you dedicate a whole day to it!

Once the edge has been reformed, it is refined gradually using more and more finer grits. We push the process depending on the use of the tool and the level of perfection that we seek. For sharpening garden tools, such as pruning shears and pruning hooks, a pass over a medium coarse abrasive then on a fine abrasive surface is generally sufficient. You can start with two water stones, one large and one fine grain. But it's a safe bet that you will quickly want to have access to more efficient and more versatile equipment.

Electric Sharpening Equipment

Electrical equipment is numerous and generally inexpensive. It allows you to work quickly.

Grinder

A portable grinder quickly removes a lot of metal. It can occasionally be useful, especially if you want to modify the shape of a tool or restore the bite to a coarse blade, like the blades of a lawnmower for example.

However, be careful with the rise in temperature on an abrasive surface rotating rapidly. Heat can change the temper of the blade and affect its hardness. Frequently dip the blade in water to avoid overheating the metal.

TWO-BEVEL BLADE

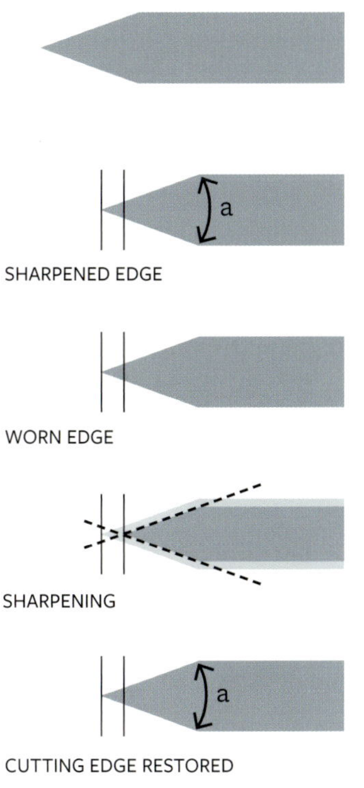

◄ *Here are two complementary sharpening machines. The one on the left is a belt sander that allows you to quickly remove a lot of metal, at the risk of overheating the steel. The one on the right is a water mill that rotates slowly for finer work.*

▼ *Here is a set of synthetic stones. The little ones of various shapes are suitable for sharpening curved blades, like gouges.*

Belt Sander

A simple belt sander, with an 80-grit belt, can be used to repair damaged tools. If, for example, a cutting edge has a large indentation following impact with a stone, a few passages on the belt can rough out the work. As with the grinder, avoid overheating.

Grinding/Sharpening Wheels

There are many pieces of equipment with a grinding wheel that can remove a lot of metal quickly. Again, if the wheel is dry and spinning quickly, be careful of overheating the metal. This type of equipment is only suitable for the initial roughing of a damaged blade. It is not used for daily maintenance of well-kept blades.

Water Grinding Wheel

The millstones (formerly natural, nowadays synthetic) soak in a tray of water. A slow rotation avoids a rise in temperature. Depending upon their grit, they achieve fine sharpening.

Manual Equipment

Sandpaper

For resharpening long blades, planers in particular, sandpaper glued to a perfectly flat surface, such as a plywood board, can do great work.

Synthetic Stones

Nowadays, excellent synthetic stones can go up to 8,000 grit. Having a set of three stones with coarse, medium and fine grains is ideal for quick and efficient daily sharpening. Electric equipment is only used on damaged blades.

Natural Stones

As a nature lover, I obviously prefer this type of stone, whether it comes from Arkansas, the Pyrenees or elsewhere, even if it is not necessarily more effective than synthetic stones for everyday use.

Japanese Natural Stones

This is not the same as sharpening garden tools, but it is more than likely that you will be working with wood on a regular basis and will need tools that cut to perfection. Those who are passionate about this discipline can investigate Japanese stones. They are the best stones in the world. Almost all Japanese quarries closed down almost a century ago and the stones that are offered for sale today are from old stock. Their price can be astronomical. But if one is looking for perfection in one's tools, one becomes interested in these stones that can achieve levels of sharpness that most craftsmen don't even know about.

Water for Stones

Most stones are used wet, sometimes with oil according to their composition. The liquid prevents pores from clogging. By mixing with metal particles, it also makes an abrasive paste that contributes to the sharpening effect.

Leather Strop

It is useful to have a leather strop, a strip of leather usually glued to a wooden support. A strop is best used with an abrasive paste. It provides the final finish after successive passes over the stones and gives a mirror-like polish.

When cutting leather, you need an extremely sharp slicer. A passage on the leather strop before each cut allows the blade to maintain its full effectiveness.

The Sharpening Process

Sharpening is not complicated in itself. You need to know how to choose the right abrasive materials according to the nature of the work to be done and work progressively with finer and finer grains.

Holding the tool correctly on the abrasive support at the angle of the bevel is important. Beginners often neglect this point and, without realising, they change the angle. There are guides that keep it at the correct angle. These guides are a little restrictive and more suitable for woodworking tools than for garden tools that often have curved blades. You have to get used to holding the tool between your fingers and ensure that the flat of the bevel is in full contact with the abrasive surface.

▼ *At the risk of surprising those who have not yet experienced manual work, sharpening can become a pleasure, passion, the expression of a quest for perfection!*

One mistake to avoid is to give the bevel a curved profile, because we then modify the angle at the cutting edge and the tool will not cut well. The bevel must remain very flat until its finest point. Extremely careful sharpening results in a blade with a cutting edge of only a few microns. The precision required means that the sharpening equipment should be set up in a well-lit area of the workshop. A magnifying glass is a useful accessory for checking the straightness of the resulting edge.

Safety

When sharpening on a grinding wheel or disc, the tool must be held very firmly to avoid losing control due to the speed of the rotation or getting it stuck between the disc and the machine casing. You must remain on high alert throughout the operation. Avoid long sleeves and scarves that could get stuck in the machine. Wear eye protection and gloves. There are cut-resistant gloves that allow a good grip on the tool; we highly recommend them to you.

Generally speaking, people don't use very sharp tools anymore. In the kitchen, for example, knives are no longer very sharp. When handling sharp tools such as razors, care must be taken. Rather than forbidding their use, inform your children of the risks: prevention is the best strategy.

Maintenance of Wooden Tools

The wooden parts of tools, especially handles, will benefit from a coating once a quarter, or failing that before wintering, with a solution made of half turpentine, half linseed oil. Linseed oil alone lacks fluidity and forms a film on the surface of the wood. To help it penetrate, it can be heated, but be careful not to bring it to a boil and be vigilant to prevent fire or burns. I did this for my boat and it acquired a beautiful amber tint and resisted years of bad weather.

A Good Workshop

To conclude this theme of tools, one thing is obvious: on any self-respecting farm, a good workshop is essential. Having your tools at your disposal at all times, properly stored and in perfect condition, saves a lot of time when carrying out repairs and maintenance work. If you like to tinker, it will become your haven. At Bec Hellouin Farm, we now have several workshops, dedicated to metalwork, woodwork or carpentry. The barn we built in 2017 has become my secret kingdom: to tell the truth, Shanti and Fénoua, our youngest daughters, have taken over one floor to create their own refuge and we sometimes make small pieces of furniture or objects together to decorate it. Sundays, the day of rest, are often devoted to small DIY projects carried out at a peaceful pace.

▲ *Once sharpened, pruning shear blades are passed over a leather strop to perfect the work. We don't have to go up to this level of finish, but with a quality tool thus sharpened, we work better and faster.*

◄ *In this workshop, entirely built by the farm team, we have everything to make garden tools, furniture, and why not a cart, a new building or boat?*

▲ *Vegetable oil is applied to pruning shears, here camellia oil.*

XVI

CREATING A MICROFARM

◂ *The farmers around us sometimes find it difficult to understand our approach, but we meet more and more of them whose confidence in 'progress' is cracking, who admit that they do not want to eat their produce. The neo-ruralists who come to the land can contribute to moving the lines of the agricultural world, if their farms are convincing. One can be vigilant about the economic balance of one's farm while fully integrating environmental, aesthetic, and even poetic and spiritual criteria.*

12

BECOMING A FARMER

History shows abundantly that only those peoples who have known how to live in happy harmony with their soil have remained prosperous and peaceful.

Philippe Desbrosses[53]

The upcoming chapters are aimed at people wanting to become farmers. Amateur gardeners may be interested in certain aspects as well.

Making the Land your Profession

Being a farmer is a job like no other. Taking up this profession transforms many aspects of our lives. There are real constraints, including for the people around us, and the pay is often low, especially if we compare it to the number of hours worked, which is better not to count! What drives us to make the demanding choice to work the land is not the attraction of a big salary, nor high social status, nor leisure... We have given up all of this.

So what makes the desire, to change one's life, so irresistible to the point of choosing such a restrictive job? Probably the fact that we only have one life, that it passes quickly and that we aspire for it to be full and connected, to have meaning. To be a farmer is to take care of life, to contribute to the well-being of your community, to have an influence, however small, on just about every major issue: health, employment, landscapes, preservation of the biosphere, climate, world hunger... We know why we get up in the morning!

Becoming a farmer is therefore a lifestyle choice and a decision that we do not take lightly.

▾ *Climatic hazards and pests challenge us, but we seek to do better every season. The quest for high quality is a powerful force.*

People who are 'Not from the Farming World'

It is likely that you are, like us, not from the farming world. Your set-up will be qualified as 'outside of family context'. Being a stranger to the agricultural world is an additional handicap compared to the person who takes over the family farm. You have everything to learn and will often be looked at with scepticism – since the wave of return to land folks of the 1970s when rural people saw a parade of idealists who don't go the distance! But choosing to farm out of passion also has some advantages: you benefit from a fresh, non-prescriptive eye, and no family circle will reproach you for changing practices. We have often heard farmers confide in us about the difficulties of transforming a farm on which previous generations sometimes still live.

Don't be too Naive

Being a farmer is more than a job... but it is still a job! One of the most difficult jobs there is.

Let's get to the point: farmers have the highest suicide rate of any profession (in France).[54] In 2016, a farmer committed suicide every two days.[55] Loneliness, remuneration that does not allow a decent living, exhaustion... It is easy to dream about the bucolic

► *Books and videos on gardening are becoming numerous. They are valuable allies; however they are not, in themselves, a complete training. We cannot overstate the need for long-term internships with professional market gardeners.*

images of the countryside in summer, to fantasise about returning to the land, but it is better to consider all the aspects of this life choice, including its dark side, before embarking on it.

At Bec Hellouin Farm, we hit rock bottom many times. This adventure led us both to hospital for illnesses linked to stress and burn-out. It has also been difficult for two main reasons: **our lack of initial training** and **the lack of reference points in terms of ecoculture.** We hope that this manual will go some way towards overcoming these obstacles. **When people ask us for advice, we tell them without a shadow of a doubt: take the time to train well!**

Having worked in many other occupations, we were amazed at the administrative burden that is now imposed on farmers.

If it can reassure you: we are a thousand times more passionate today about our profession than when we first approached it and we do not regret embarking on this adventure, despite its difficulties. But if I had to do it again, instead of plunging in, we would progress step by step, while taking the time to properly prepare each new phase and enjoy every moment of life.

Train Yourself

Do you know of a single profession that can be mastered in a few months? Most take years of study. Our friend Jean-Martin Fortier, the famous Quebec market gardener, states: "It takes 20,000 hours to become a market gardener." After four to five years of hard work, we agree!

In recent years we have watched as several 'urban people' have changed their lives to get

Perrine's point of view

I have had the opportunity to consult on numerous agricultural projects, as a regional advisor in charge of organic farming for Haute-Normandie, and as a participant in the activities of our farm. I learned how complex the agricultural world is. It has its codes, its customs, its rules and its laws! These were created to protect this status and avoid too much degradation of the environment for projects or unserious people, in particular in order to limit the loss of agricultural land. Creating a farm from an administrative and legal perspective is much more complex than creating an ordinary business. Those who have experience as a business manager might underestimate the difficulty of the journey ahead. Beware of that! Even for people from the agricultural world, the journey of creating a farm is a headache. It's necessary to become familiar with the codes, uses and rules, and above all know that the creative journey is often long (around eighteen months to two years); you must therefore be patient. Often, new farmers carry a lot of emotion and the slightest annoyance or difficulty brings them turmoil. You have to accept that not everyone will be immediately and completely enthusiastic about your project; some will not even understand it and may have harsh words. Don't let that discourage you; don't get angry either, because that won't help. Tell yourself that they may not be at your level of understanding of the issues facing our planet, so have compassion for them, explain and re-explain. At each difficulty of our journey (including once established), when we were able to discuss and explain, we realised how wide the doors open (admittedly, sometimes slowly) when we manage to establish a respectful dialogue.

back to the land and quickly proclaim themselves to be 'experts in market gardening' and carry out consulting activities, even though they had never really been farmers. This indicates a certain contempt for the profession, as if there was no need to learn it...

Take the necessary time to train properly with professional farmers and over time in the field. Certainly, we arrived at farming rich with the experience of a previous life, even if it had nothing to do with farming. The skills and maturity acquired will be valuable in one way or another, but this does not take the place of good training.

There is a paradox: agricultural education does not yet prepare one for new forms of agriculture, the subject of this manual. Teachers are not familiar with it, even though in France, there is a strong movement toward microfarms and permaculture that is changing the way agriculture is taught. At the end of 2017, the institutions in charge of agriculture tell us that 80% of market gardening projects follow a permaculture farming model.[56] While waiting for permaculture and ecoculture education to become widespread within the agricultural world, you will have to find your own path. We suggest taking several comprehensive training courses in different contexts.

Create your own Training Syllabus

Here is a complete training course outline, if you have no agricultural background.

1. **Test yourself by doing practical training with farmers.** Not just one try, nor a week in the summer! These courses will be the opportunity to confront the reality of daily life on a farm.
2. **Complete training in organic farming,** choosing a course that is as close as possible to your project. It is possible to follow certain training courses continuously or by correspondence. A diploma in organic market gardening gives good foundations during a year of study, within which we carry out several weeks of practical courses. In addition, it is one of the degree programmes which give you access to financial aid for new farms.
3. **Take a Permaculture Design Course** (PDC), as well as short training courses on themes you wish to explore further.
4. Take training in ecoculture, such as those that we offer at Bec Hellouin Farm. This is professional training for farmers, with a permacultural approach and the techniques of microfarming. These training courses are extremely rare to date, but they can be replaced by this manual which contains much more information than we can offer in a few weeks of lessons.
5. **Do long-term practical training courses** over several months, ideally a year on farms that are close to your aspirations.

Some people are reluctant to volunteer on a farm. This seems to us to be a miscalculation. Remember that the time the farmer spends training you is well worth the work you do for them in return. It is better to make mistakes at home than for others. The time you spend on training will save you a lot of unnecessary expenses later on and will make you more productive sooner.

The model at our farm requires little material investment, in any case less than other forms of agricultural installations, but it does require a lot more knowledge and skills. You have to invest in yourself. Training takes time and money. There is no miracle; it is unavoidable.

Institutions of the Agricultural World

If you were not born on a farm, your time training will help you learn to navigate the galaxy of institutions in the agricultural world. In France the profession of farmer is very supervised, the legislation dense, and administrative tasks cumbersome considering the low pay of the profession. Organic farmers are not the best off, far from it, because the organic label adds an additional level of complexity. You might as well approach it positively and understand these institutions and administrations: you will meet people who are passionate

▲ *Exploring new ways to produce our food is an exciting cognitive adventure. We enjoy drawing information from various sources, one day immersed in reading testimonials from pre-industrial agriculture, another from the most recent scientific publications.*

▼ *We also appreciate the diversity of encounters that being a neo-farmer brings. In this photo are the permanent team of the farm, our family and trainees, farmers in training and students who have come from Sciences Po or agronomy schools to do their end-of-studies thesis at Bec Hellouin Farm.*

about it, and sincerely eager to support your efforts. The distrust we sense is caused by the dreamers who are far removed from the realities of the agricultural profession. It's up to you to demonstrate your credibility. And if some doors close because the approach is atypical, persevere: others will open up. The agricultural world is changing and the projects that best integrate environmental and societal dimensions are the projects of the future. Do not hesitate to ask an experienced contact person (inspector or director), who will be more aware of trends.

The diversity of tasks to accomplish and skills to acquire for new farms can make your head spin... Things calm down when we plan the transition over several years.

Status

In France, you can become a farmer according to different statuses. It is important to know them and opt for the one that best suits your project. Your status will depend, among other things, on your active agricultural area and the type of production chosen.[57] Recently, it has been possible to obtain the status of primary farmer depending on the number of hours worked per year, which opens the door to installations on very small areas that were penalised until now.

Social Coverage

In France, the status of primary farmer gives you access to managed social protection by the Mutualité Sociale Agricole (MSA). It works pretty much like other funds: social charges are indexed to your income. Generally speaking, contributions are not very high, but the benefits and pensions are not either.

Taxation

In France, you can opt for different tax statuses. The first choice is fixed, calculated according to your land use and type of production; the second is actual, calculated according to your profit (receipts minus expenses). Many people think that flat rate taxation is better but it is often a bad choice because this means you will be taxed every year, whether you make a profit or earn nothing at all. In addition, you will not be able to depreciate your investments. However, during the first few years following a set-up, we generally invest a lot but gain nothing... Taxation on actual gains takes into account the investments will depreciate over five, ten or twenty years, to reduce your taxes over the years when you get your head above water.

The second choice concerns VAT. You can opt to adjust liability or not. There again, it is often wise to opt for insurance VAT liability, which will allow you to deduct VAT from your investments and purchases, if you decide to invest significantly.

Setting up a microfarm must be approached like the creation of a complex business, despite its small size. Amateurism is generally heavily sanctioned. As you can see, there is enough to keep you busy for a few years! Is this really a problem? When we embrace the profession of farmer, we make a long-term commitment. Once you have established a solid foundation and got your bearings, you will be able to enjoy your farm for a long time, and your descendants will too!

▼ *A person lying in the grass savours the beauty of the Bec valley... It's a safe bet that this is not a member of our permanent team, because the work here is intense and rarely leaves us the time to take a break. But we can appreciate the surrounding beauty while staying productive!*

SKILLS TO LEARN AND TASKS TO ACCOMPLISH WHEN CREATING A FARM
FINANCING
LOANS
SUBSIDIES (INCLUDING THE YOUNG FARMER GRANT)
PARTICIPATORY FINANCING
PROVISIONAL OPERATING BUDGET OVER FIVE YEARS
FINANCIAL PLANNING
ESTIMATED INVESTMENT BUDGET
LEGISLATION
LEGISLATION RELATING TO THE PROFESSION
FARMER'S STATUS
KNOWLEDGE OF THE STAGES OF THE INSTALLATION PROCESS
REGULATIONS PER TYPE OF PRODUCTION
CAP AID
SOCIAL SECURITY COVERAGE
MSA MEMBERSHIP
CONTRIBUTIONS
TAXATION
TAX STATUS
INSTITUTIONS
RELATIONS WITH AGRICULTURAL INSTITUTIONS
ORGANIC FARMING
AB SPECIFICATIONS
CERTIFYING BODY
ORGANIC REGULATION BY TYPE OF PRODUCTION
INSURANCE
SITE AND EQUIPMENT INSURANCE
LIABILITY
BUILDING
INVESTMENTS
EQUIPMENT
SUPPLIERS
PRODUCTION
PLANNING
CROP MANAGEMENT
HARVESTS
TREES
ANIMALS
PROCESSING
SPECIFIC LEGISLATION
EQUIPMENT
WORKSHOP
PACKAGING
MARKETING
SALES
CUSTOMER DEVELOPMENT
POINT(S) OF SALE
VEHICLE
COMMUNICATION
VISUAL IDENTITY
WEBSITE
GRAPHIC CHARTER
MANAGEMENT
ACCOUNTING
BILLING
MAINTENANCE
EQUIPMENT
BUILDING
LAND
ACQUISITION OF LAND
HOW TO PRESENT YOUR PROJECT

◄ *Not owning land makes creating a new farm more complex. This situation, however, offers the freedom to truly choose a place to put down roots, according to your tastes and professional choices. The permaculture approach described in Part III of this manual allows you to match the farm desired to the environment, a guarantee of success.*

13

IN SEARCH OF A FARM

Mother Earth knows her child, my hands pull plants, food and shelter from the soil. Father Sky knows the face of his children, lifted up to him in wind, rain, fog and clouds.

Anna Lee Walters[58]

Defining your Project

There is one essential prerequisite for starting a farm: defining your project. You wouldn't expect an entrepreneur to start a business with a vague idea. The same applies to your farm.

Once your project is well defined and all aspects are considered, present it in a credible business plan. You will have to defend your project before various bodies: land and rural development (SAFER, in France), chamber of agriculture, local authorities, the bank and more. Your business plan will be useful at every turn. Take the time to create a quality document that is a testament to your seriousness and professionalism.

An investment budget, a five-year forecast budget and a financing plan are essential. Your contacts will look at them carefully and read between the lines to identify if you are out of touch with the realities of the profession or well prepared.

An installation advisor can guide you along the way. You can also subscribe to the 'young farmers' course and benefit from more in-depth support. [59, 60, 61, 62, 63, 64]

▼ *The presence of the Bec stream was a determining factor in our choice to acquire the 6,500 metre plot of land where the farm was built.*

Setting Up in the Countryside

We will now discuss the creation of a microfarm in a rural area. In the next chapter we will explore city microfarms. Choice of location determines so much for the future of a farm that this stage deserves attention to detail. Your ongoing training in permaculture will allow you an in-depth look at site potential.

Where to set up? Unless you already have a site, this question arises first. Finding 1 hectare in the countryside or on the outskirts of towns is not very difficult (apart from cost decisions). But choosing **your future farm** requires attention to many parameters: the beauty of the place, its climatic conditions, the composition of the soil, the presence of a river or a forest, the nature and condition of existing buildings, surrounding development, the dynamics of the area, water resources... Refer to the questionnaire in volume one.

Parents will consider the proximity of a good school; farmers will look for places to sell their produce. There is no ideal place, of course, and the final choice will result from the pros and cons between many parameters. We have to make compromises, but if the gap is too big between a location and your dreams, you will have to work a lot and invest a lot to reduce the difference... It is therefore better to devote more time to your quest rather than rush. Besides, visiting various places will help you hone your judgement and, on the day when the farm of your dreams appears, you will be able to decide in just a few hours.

Organisations that Help Access Land

The agricultural land market is specific. Any agricultural 'zoned' land transaction in France is managed by the **Société d'Aménagement Foncier et Rural** (SAFER), that lists farms for sale. If you wish to acquire agricultural land, you have to present a business plan of your project. **SAFER** decides in a committee of land allocation. Other countries may have different rules.

The **Terre de Liens** association can help you acquire agricultural land. You will not be the owner, only a tenant with a lease for the duration of your career.

"You do not inherit the land from your ancestors, you borrow it from your children."

Attributed to Saint-Exupéry

Buying a House and Land

Those who aspire to settle down as farmers are most often faced with a dual problem: needing land and a place to live. The good news is that many houses in rural areas are sold with a land area that authorises the creation of a microfarm (usually the surface area of the garden only marginally impacts the value of the property). All provincial real estate agencies propose this type of property, often with some outbuildings in addition to the dwelling. The acquisition of a house with one hectare of land or more resolves both problems at once, and a third benefit is that this type of land is, most often, not considered by administration to be agricultural. The price is not regulated by SAFER. Its sale is regulation free: no need to go through any administrative procedures that are often long and complex. The downside is the cost will certainly be higher than agricultural property.

Once we have land and produce agricultural products, even if these lands did not previously have an agricultural destination, it becomes possible to gain the status of farmer, or at least as a solidarity contributor to the MSA, which gives the right to sell produce. It is this simple path that Perrine and I followed for Bec Hellouin Farm, with the purchase of an old cottage nestled in a 6,500 square metre orchard.

If you can, we advise you to live on your microfarm: you will benefit more from it, your family living environment will conform to your aspirations, there will be no commute and you can easily perform daily farm tasks. There are some constraints: the distance of schools, the intrusive aspect of too many visitors... A good design is necessary.

▼ *Émilie Sage is a shepherdess on the island d'Yeu, in Vendée. I knew her as a child when this island was our homeport. As a teenager, she spent a year on board my school ship,* Fleur de Lampaul. *As a young adult, Émilie trained in permaculture on our farm; we are proud and enjoy following her journey! On her farm, Émilie and her companion raise Solognote sheep, berries and aromatic plants, and offer training. It is a great joy to see bright and courageous young people create exemplary farms and generously share their knowledge.*

The Farm of your Dreams

There you have it, the land is acquired! Let's imagine an ideal scenario: you were able to buy back an old farm, you have a residential house and some old barns or outbuildings, and even a cellar, a pond, a well... This will save you a lot of time and silver. You also benefit from a stream, or in the absence of one, a beautiful pond: the dream! And if you can acquire a few additional hectares of wood/woodland, don't hesitate for a second!

Organic Certification

As soon as your domain is acquired, we advise you to get organic certification. If your land was used for conventional farming, you will have to go through a period of conversion before being able to sell products with an organic label. But if you buy grassland or unexploited land, then you may benefit from an exemption and be certified organic without a conversion period – check the regulations for your area and country. Wait to be certified before turning over the soil. In France, grasslands generally qualify for an exemption for organic certification, but the controller must be able to note the presence of grass! On the other hand, in order to preserve the environment, meadowland is under surveillance and ploughing cannot be done without authorisation. So find out the status quickly from your local land management officials.

It is up to you to choose a certification body, and you are responsible for paying for their services. Some communities contribute to certification costs.

Set Up Aid

By obtaining farmer status, you can possibly benefit from financial aid.

In the EU, if you have agricultural credentials or use the acquired experience route, you can receive financial aid called 'young farmer endowment', depending on numerous criteria and the region; this can range from approximately 12,000 to 38,000 euros.

Some local authorities award a subsidy for organic farms. Once installed the farmer can receive aid from the Common Agricultural Policy (CAP) and/or programmes that help organic agriculture. There are also, depending on the area, subsidies for planting hedges and other agro-environmental measures. Your consultant will be able to advise you.

◂ *The rise of microfarms can breathe new life into countryside that was largely emptied during the period of rural exodus. Each farmer contributes effectively to the local economy by providing work for craftsmen, traders, restaurateurs and tourism. Dynamic rural places generate a boom in services, community and culture.*

Transform your Land into a Microfarm

Each project is unique; however here is a summary of the main 'measures' to be taken for owning a farm, summarising the principles described throughout the chapters of this manual.

Suppose you have acquired a one hectare microfarm. If you decided to become a market gardener, to ensure you earn a living, 1,000 square metres maximum should be dedicated to vegetable crops. What to do with the 9,000 square metres remaining?

Implement the main principles of eco-culture, as described in volume one; these principles are aids in the design of a system.

Take the time to find out what the spontaneous ecosystem was before human intervention. Listen deeply to nature: what does it want? Try to put yourself at the service of your project, to harmonise your dream with the dream of the Earth. The Earth wants to feed us; its mission is to support the existence of living beings. Have faith in the Earth - by entering into a co-creation process with her, you will move forward much faster and surely!

Prioritise Fruit Trees!

In almost all latitudes, the main priority will be to plant many fruit trees. Their growth will help absorb carbon released by the creation of the microfarm and the activities of its occupants. They will provide food, biomass, humus, shelter for fauna and flora, create microclimates and more.

We favour fruit trees (including nuts), because the nutritional function to the microfarm is so important. They will also provide additional benefits like a little firewood, or even wood for crafts or construction, but the latter uses will be better filled by trees growing in nearby woods or forests.

Fruit trees can be introduced in a variety of forms, allowing hundreds of trees of many species and varieties to be planted:

- Ideally, it is desirable to surround the middle farm with a fruit hedge that fulfils many functions.
- A **forest garden**, even if small, will naturally take root on the dominant windward side. The forest garden will produce a great amount of diversity and can serve as a place for animals.
- A **meadow orchard**, with tall or half-stem trees, will allow you to harvest fruit, and shelter animals.
- A **market garden** dedicated to vegetable crops will benefit from being managed in agroforestry.
- **Trees trained** on walls will benefit from the accumulated heat of the stone.

These hundreds of fruit trees give a microfarm its oasis character. When they mature, the canopy will cover most of this small area.

In our vision, the vegetable crops, which are mostly annual, may represent the main contribution in terms of turnover, but they remain secondary from the point of view of the ecology of the place.

▲ *When looking for agricultural land, it is better to be demanding about its criteria. The presence of a watering point or a river is a definite asset. It is often preferable to look for the rare pearl for a long time, than rush into an acquisition that will then require long and expensive work.*

Recreate the Link between Livestock and Crops

We have mentioned the animals that can be raised in a microfarm: sheep and goats, poultry, pigs, rabbits, a donkey or a pony.

A small cow or a small draught horse is a great addition, but be aware that they require 1 hectare of meadow to feed, at least 12,000 square metres, or you'll need to buy hay in. Avoid overgrazing: the tendency to want to breed all kinds of animals (we fell in love with a goat and its kid at the market!) is likely to push farmers to accommodate too many tenants.

Animal-crop synergy has many benefits. The presence of animals facilitates the closing of cycles: in addition to grass, they will eat the remains of crops, vegetable waste, hedge trimmings, reeds from ponds... Their excrement will make an excellent compost for trees and crops.

▶ *Our land was not a farm – we had to build almost all of the buildings, causing a lot of fatigue, expense and worry.*

Field Crops

If market gardening is the main operation, it may be interesting to dedicate 1,000-4,000 square metres to a plot of land cultivated with animal traction, in order to complete the diversified vegetable production of the permanent beds with field vegetables: potatoes, squash, leeks, cabbage, root vegetables...

Grains

It is possible to grow grain in a garden (see Chapter 9, p.93) and to build a bread oven.

Beekeeping

Beehives naturally find their place on a microfarm, for honey and pollination... without neglecting wild bees, by taking measures that promote biodiversity.

Aquaculture

It is possible to raise fish in ponds, for example carp that require no inputs, or other hardy local fish. In addition to the enjoyment of living aquatic environments, the fish can be put to good use: if not for human consumption, they can be sold to restock ponds in the region.

Rivers, Ponds and Bodies of Water

As mentioned in the search for a microfarm location, we highly recommend choosing a location near a river. The presence of water is an invaluable asset. It will irrigate crops (if it is not polluted), water the animals, run a microturbine and more.[65]

In the absence of a river, we can create ponds. One does not prevent the other: our farm is crossed by the Bec stream, but we dug 25 ponds and water points for crops and animals. The pleasure of a river is an additional asset, for swimming and the beauty it provides.

Ponds contribute to the richness of a site's biodiversity. They are a source of biomass: silt that settles at the bottom is a concentrate of nutrients, especially if fish are present; reeds can be used as mulch, animal feed and also human food when young; fast-growing aquatic plants can be planted and used as mulch.

◀ *An eco-built house offers many advantages. It can produce its own energy and recover the water and waste of its inhabitants.*

In addition to aquaculture, mentioned above, water features can be dedicated to the cultivation of spirulina, a micro-algae very rich in proteins.

Eco-Habitat

To be consistent with the ethics of permaculture, it is obviously desirable that any dwelling is designed in an ecological way, or at least transformed to approach it. A bioclimatic design with excellent insulation saves or can even eliminate heating costs. If there is a heat source, it can be provided by a cookstove that uses wood from the property. It can also heat radiators. Once you have experienced the charm of a good woodstove, designed to ensure the best possible combustion of smoke, it becomes difficult to do without!

Attaching a greenhouse to the south side (in the northern hemisphere) of the house will provide a frost-proof greenhouse in winter and save two months of heating per year. And the young plants will be well tended!

Our dream at Bec Hellouin is to transform our thatched cottage into an off-grid house, not connected to water and electricity networks - a house that collects its own water and energy within a microfarm would be utopian.

The house offers various resources:

- o Rainwater, collected by gutters, carefully stored in tanks for various uses (domestic uses other than drinking water, irrigation), while grey water can be used for irrigation (a basin of phyto-purification can be positioned ahead of the irrigation system);
- o Heat leaving the home will generate microclimates for the benefit of sensitive crops;
- o Organic matter (meal leftovers) and ashes from the wood stove will be recovered (feeding chickens or pigs, compost...);
- o Dry toilets complete the mineral cycles.

This illustrates the unsuspected potential of a very small piece of land. Obviously, not everything can be accomplished at once; phasing makes it possible to spread work over time, by prioritising things. The forecasted budget reflects the choices and trade-offs made by project leaders.

The small size of a microfarm allows it to function in all kinds of environments, including in the heart of cities. A number of topics covered in this chapter can work in an urban setting.

▼ *Large animals can be raised in a small area. The economic viability of such a project is only possible if we process the production ourselves in order to create value on the farm.*

la Ferme de NATURE & DÉCOUVERTES
S
E
O
N
NATURE & DECOUVERTES

◄ We had the pleasure of working on this adventure with landscaper Rémi Algis and project leader Gilles Degroote. This urban microfarm boasts market garden crops, grown in both flat and rounded beds alongside a forest garden. It is open to the public and offers training sessions.

14

CITY FARMING: URBAN AND SUBURBAN AGRICULTURE

Our environment could be much more pleasant to live in. We could reclaim our streets, without having to demolish or rebuild everything. For this to happen, we, the residents of the street, can no longer turn our backs on it and lock ourselves behind screens, walls and solid doors. If we take to the streets, we can initiate very different processes and consider that the street is also our business.

Nicolas Soulier[66]

The development of urban agriculture is a hot topic. On every continent, many local elected officials aspire to bring agriculture to the heart of their city. This trend, however, is up against the scarcity and high cost of land, as well as problems with pollution. The permaculture approach developed at Bec Hellouin is of great interest because a lot of food is grown in a small area.

Improving the Resilience of Megacities

The growth of cities is a well-established trend. More than half of the world's population lives in urban areas. There are a growing number of megacities with populations of more than 10 million, several exceeding 20 million. The growth of these giant cities has been made possible by the immoderate use of cheap fossil fuels. What will become of them in a few years, when these energies become rare and expensive? If I had money to invest, I would certainly not invest it in urban housing! Cities are very fragile superorganisms because they depend on the outside world for almost all their supplies. Our European capitals have food stocks for only a few days. If their supplies were to be interrupted, famine would strike. Megacities have no resilience for water, energy or food.

In our book, *Miraculous Abundance*, we dealt with this subject, evoking the fact that the energy descent will bring about the decline, or complete collapse, of the individual automobile civilisation.[67] A drastic reduction in the number of cars would free up a lot of space in the heart of cities, which would allow the reclamation of concrete and asphalt by food gardens, for the benefit of local communities.

Microfarms with Multiple Functions

In the city, microfarms relocalise food production, without generating any dependence on fossil fuels; it's their main advantage. This new type of farm produces much more than food though, and the other functions are particularly interesting in densely populated areas. The following list is not exhaustive. Leaders of urban and suburban installations can point to these functions in their discussions with municipal authorities.

Local organic food: by producing quality food locally, urban microfarms eliminate long, energy-intensive supply chains, the need to freeze or transform fruits and vegetables by industrial processes, and transit through large distribution markets.

Health: garden produce can be consumed as soon as it is harvested by local residents. The freshness of local garden production is incomparable. Their nutrients are preserved to the benefit of consumer health. Not to mention the positive effect of gardening for those who put their hands in the ground!

▲ *Gilles Degroote was a consultant before becoming an urban farmer. He devoted several years to training and preparation for his project. The quality of his preparation aroused the trust and support of stakeholders as diverse as the leaders of Nature et Decouvertes, managers of the Versailles estate, and the city's elected officials.*

▼ *The construction equipment had barely left the plot when vegetable production began. An urban microfarm is constantly under the scrutiny of local residents and elected officials; it must quickly demonstrate its relevance.*

Food security: cities become less dependent on long-distance supply routes that are energy-intensive and vulnerable. In case of societal collapse or geo-unrest, the vital needs of the inhabitants can be ensured locally. There will need to be many microfarms, however, to ensure real food security in the event of a major crisis.

Renaturating the urban environment: greening cities, stocking them with flower gardens, and adorning them with lush food gardens is clearly beneficial for everyone. Studies show that if hospital patients have a view of a tree, the average duration of the hospital stay is significantly reduced.[68] The greening of cities is a very effective way to reduce vandalism. When will there be 'edible cities', for the well-being of inhabitants?

Microclimates: the presence of plants, especially trees, tempers the heat of the sun and moderates the wind while humidifying the air. We complain about 'unbreathable' air in the city, but never in the countryside!

Biodiversity: various studies have demonstrated the role of urban vegetable gardens as biodiversity reserves.

Depollution: the presence of plants and arable soil cleans the air and water.

Soil remediation/cleaning: polluted soil must be restored before becoming productive again.

Energy autonomy: it takes 10-12 food calories to provide 1 food calorie on our plate. When production is relocalised, energy efficiency in agriculture is increased tenfold.

Reduction of transport: more food production in urban areas means less road transport. If widespread, cities could be rid of commercial vehicles.

Job creation: urban farms create opportunities for residents to start businesses. However, the question of the status of farmers in the city needs to be considered: it is likely that the legislation will have to evolve.

Social link: citizen initiatives to transform urban spaces, such as 'Incredible Edible', are powerfully unifying. Collection of organic waste, shared workload, the distribution of CSA baskets are all social occasions.

Solidarity: solidarity networks can be set up around a microfarm. This can be a factor towards integration. Our AMAP programme (Maintaining Small Scale Family Farming) in Paris provided solidarity baskets and cooking classes to poor neighbourhoods.

Education: an urban farm is a place for environmental education.

Quality of life: a single microfarm can nourish a few hundred people, but many thousands more benefit from the beautification of the neighbourhood.

Social stability: a food garden reduces the suffering of populations cut off from nature.

Valorisation and recycling of urban waste: urban farms help close the organic material cycle within the city.

Production knowledge: the question of food security in cities needs to be explored.

Training: an urban farm can provide training courses that are useful for local residents.

The Economics of a Farm in the City

The economic realities of a farm in the city are not the same as rural farms. This sensitive point deserves attention.

If we only consider food production, despite the opportunity to sell directly to city residents at a good price, it is very likely that the activity will not be profitable. Indeed, an expensive urban location means operating costs are higher than those of a farm located in the countryside.

It therefore seems essential to consider all the services rendered by the farm to the city,

and to remunerate at least part of these services. This remuneration can take the form of direct or indirect aid to the farmer by the community. It is also possible to charge for certain activities: paid visits to the farm, gardening lessons, food awareness and healthy food.

We participated in a project to create a microfarm on the roof of a department store. The store managers acknowledged that offering the roof and the initial investments would not be enough to make it viable for a market gardener to earn a living: the operating costs of a rooftop farm and the time spent dealing with customers would be prohibitive. The department store benefits from the presence of a microfarm on its roof, as an attraction that brings in customers. This justifies having all or part of the market gardener's remuneration borne by the store for services rendered. That way, the farmer benefits from a decent level of income to rent accommodation in the city that is more expensive than in the countryside. A farm in town weaves together all kinds of partnerships, and a partnership is only lasting if it is fair.

An urban agriculture project must therefore integrate all services provided, and not just the agricultural production. This assumes that elected officials and the inhabitants of the city are collectively aware of the importance of these services. The first farming operations will face obstacles and smooth the way for future projects, but the trend is established, and nothing can stop it.

Soil Pollution

Vigilance about the condition of soil is necessary in cities, particularly heavy metals. Specialised laboratories analyse soil. When pollution

▼ *Creating a farm where land is scarce and expensive, and local expectations are high, is a demanding adventure that leaves no room for amateurism.*

is present, different strategies can be considered that can go as far as replacing soil or above-ground crops, making organic certification impossible. In all cases, long-term bioremediation or phytoremediation strategies are necessary.

Access to Land in the City

The rise of urban agriculture is a worldwide trend that will grow stronger in the decades to come. It is about food safety for cities, in a post-oil world.

The price of land in the city is such that dedicating it to food production seems unthinkable. Increasing land cost chased the market gardeners out of Paris at the end of the 19th century. Better to think about a new allocation of land already available. This lands may belong to local authorities, businesses, communities, associations or individuals.

Land Owned by Local Government

Local authorities generally have a number of locations that can be made available for urban agriculture projects; it may be parks, wastelands, courtyards, areas bordering railway tracks or rivers... This assumes political will and adapted urban planning regulations. As mentioned at the opening of this chapter, political will is strengthening, it is becoming *mainstream*, as we can see from the large number of elected officials, town planners and landscapers who come to visit us or train at our farm.

The cost of maintaining an urban park is generally high, similar to the cost of operating a microfarm. This consideration can influence the transition of parks towards agricultural use because the operation does not represent additional cost for the community, on the contrary.

▼ *Urban microfarms create social links. They contribute to fulfilling the need for nature among city dwellers and beautify their daily lives. The value created goes far beyond agricultural production.*

Corporate- and Government-Owned Land

Many companies have land surrounding their head office or facilities (factories, warehouses, etc.). Some have made land available to employees or a market gardener. This could change the atmosphere of the company. A business manager friend in Ukraine has created a permaculture farm for the benefit of employees.

We have a number of trainees sent by hospitals who understand the benefits of establishing therapeutic gardens for patients. So often the means are lacking, but the desire is there.

Land Owned by Communities or Associations

Urban land owned by religious organisations and solidarity associations is sometimes considerable, and all the more surprising as it is often hidden behind high walls intended to preserve the tranquillity of the occupants. It's generally of an ancient heritage, dating from the 19th century or earlier, the sustainability of which has been ensured by the stability of these institutions. Today, these communities have aged and their numbers have melted like snow in the sun. These estates are often looking for new uses for the land and the non-commercial, charitable or spiritual status of these institutions makes them favour collective uses. Many of these preserved spaces are assigned agricultural use with a strong social component. Micro farmers are increasingly in demand by religious communities wishing to integrate permaculture into their establishment.

In Cuba, I saw students working in the school's vegetable garden. In this country where precariousness is high, local resilience is achieved by creating edible spaces wherever possible. All schools have a vegetable garden and individuals cultivate tiny spaces. Our educational establishments could draw inspiration from it – some are already doing so and have created a permaculture garden.

Land Owned by Individuals

There are millions of urban gardens in France both small and large. Given the potential of micro-agriculture, each of these gardens can feed several families in the neighbourhood, or even more depending on their size. There is something deeply fascinating about the

◄ *A simple balcony can become a little oasis of greenery and help provide vegetables and aromatics for city gardeners. The symbolic significance of this food goes well beyond the volume produced. Here is Lila's balcony, in the heart of Paris. These few pots randomly placed have great value for her.*

transition of lawns that occasionally hosted children's games, or a Sunday barbecue. There is almost no investment, other than labour, to transform them into fertile oases.

The Dervaes family's food-producing oasis, in California, is well known: on 370 square metres of urban garden space, a father and his children earn $20,000 in annual turnover (more than 17,000 euros)!

Some owners of suburban homes are unemployed, retired, or have a young adult who cannot find a first job. Still others suffocate in an air conditioned office, between the computer and the coffee machine. All they need is solid training to begin to produce high-quality vegetables in little spaces. Mescluns (young salad leaves), aromatic plants and edible flowers can bring in additional income right at home.

If this transition is still seldom realised it is because awareness is low. It is not on the radar of city dwellers who are out of touch with the land. Few believe it is possible to 'become a farmer' at home, in the city! By increasing awareness of microfarming and explaining how to earn a living with just a few dozen square metres, we hope that this manual will inspire new projects.

The permaculture approach invites us to pay particular attention to urban microfarming. How can ecosystems be put to good use? What is the source of biomass that will make it possible to maintain a high level of productivity? The rise of urban farming invites us to think of the city as an ecosystem and to multiply the connections between all its stakeholders: municipality, individuals, associations, composting platform...

▼ *Below, the King's Vegetable Garden, Versailles, home to the renowned École Nationale Supérieure de Paysage. Training future landscapers to integrate the nurturing dimension of gardens.*

The example of the Dervaes family farm

In response to a food scandal in 1985, the Dervaes family set out to produce their own healthy, organic food in Pasadena, California, just minutes from downtown Los Angeles. On just 370 square metres, they now produce 400 types of vegetables, fruits, edible flowers and aromatic herbs. Their small urban farm combines abundance, productivity and efficiency. By paying particular attention to the soil and its fertility while saving water and energy, the Dervaes family produces nearly 3 tonnes of food, 1,800 eggs and 200 litres of milk each year. The production feeds a family of four and generates an annual turnover of $20,000 with the sale of fruits and vegetables. These products attract neighbours as well as great chefs. Expenses are kept to a minimum and most of the work is done by hand. These results are, however, not obtained without effort: each member of the family works on average 10 hours a day.

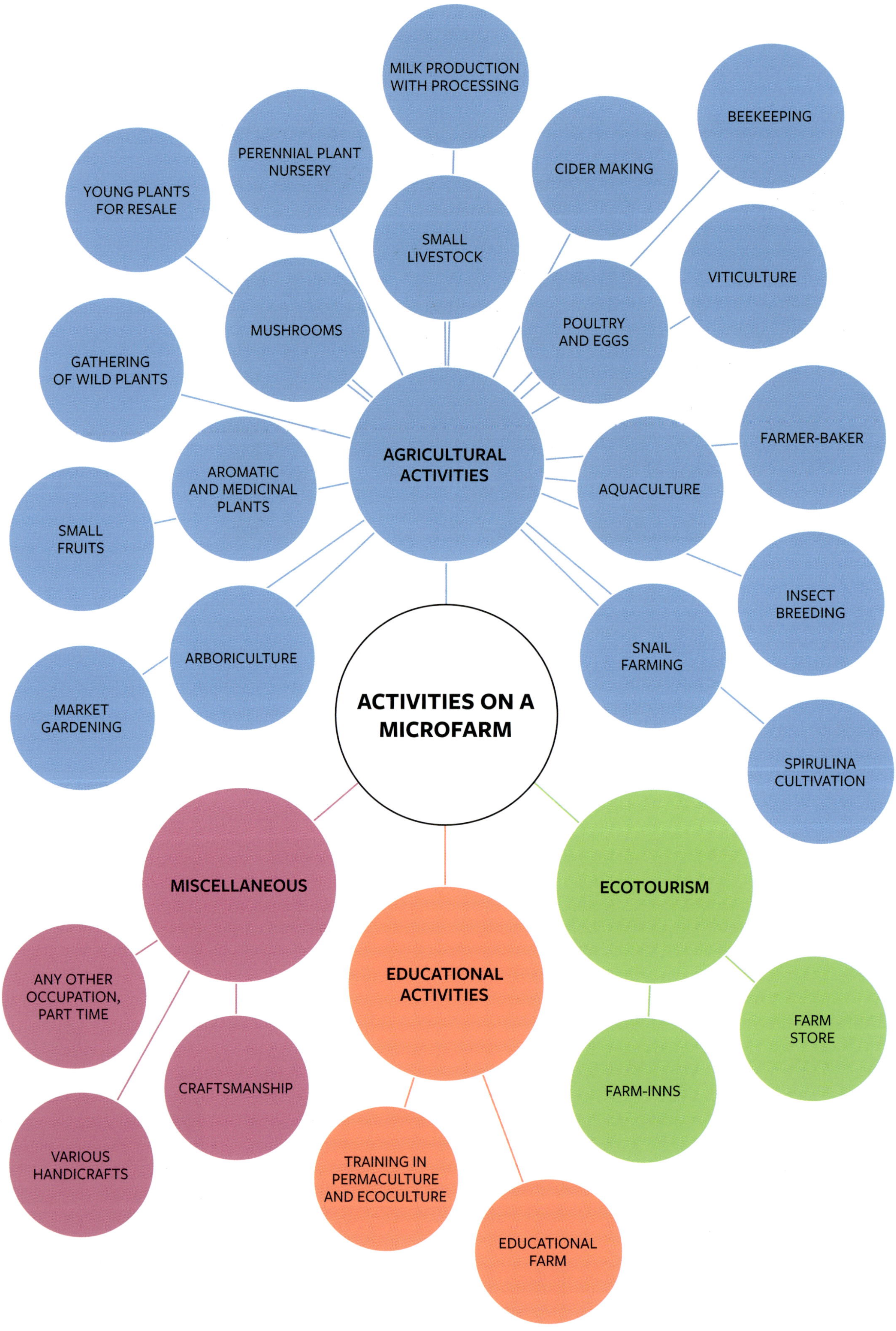
ACTIVITIES ON A MICROFARM
AGRICULTURAL ACTIVITIES
MILK PRODUCTION WITH PROCESSING
BEEKEEPING
PERENNIAL PLANT NURSERY
CIDER MAKING
YOUNG PLANTS FOR RESALE
SMALL LIVESTOCK
VITICULTURE
MUSHROOMS
POULTRY AND EGGS
GATHERING OF WILD PLANTS
FARMER-BAKER
AROMATIC AND MEDICINAL PLANTS
AQUACULTURE
SMALL FRUITS
INSECT BREEDING
ARBORICULTURE
SNAIL FARMING
MARKET GARDENING
SPIRULINA CULTIVATION
MISCELLANEOUS
ECOTOURISM
EDUCATIONAL ACTIVITIES
ANY OTHER OCCUPATION, PART TIME
CRAFTSMANSHIP
VARIOUS HANDICRAFTS
TRAINING IN PERMACULTURE AND ECOCULTURE
EDUCATIONAL FARM
FARM-INNS
FARM STORE

15

DIVERSITY FOR MICROFARMS?

The transition to a balanced social and economic system requires a corresponding change of values. Assertiveness and competition will need to give way to cooperation and social justice; expansion to conservation; material acquisition to internal growth.

Fritjof Capra[69]

▼ *A small farm and family diversifies its activities by creating high economic and societal value in a small area. At Bec Hellouin Farm, agricultural production and hospitality are closely associated.*

The preceding chapters give an idea of the potential of microfarms. The field of possibilities is only limited by your imagination! Many activities in the agricultural sector take place in a small area of a few hundred square metres to a few hectares. Here is a non-exhaustive list of possible activities to inspire you, even if you can't achieve them all.

Diversity, a Factor of Success

To get the most out of a small farm, it is wise to incorporate different functions. The advantages of bringing diverse activities together in one place are as follows.

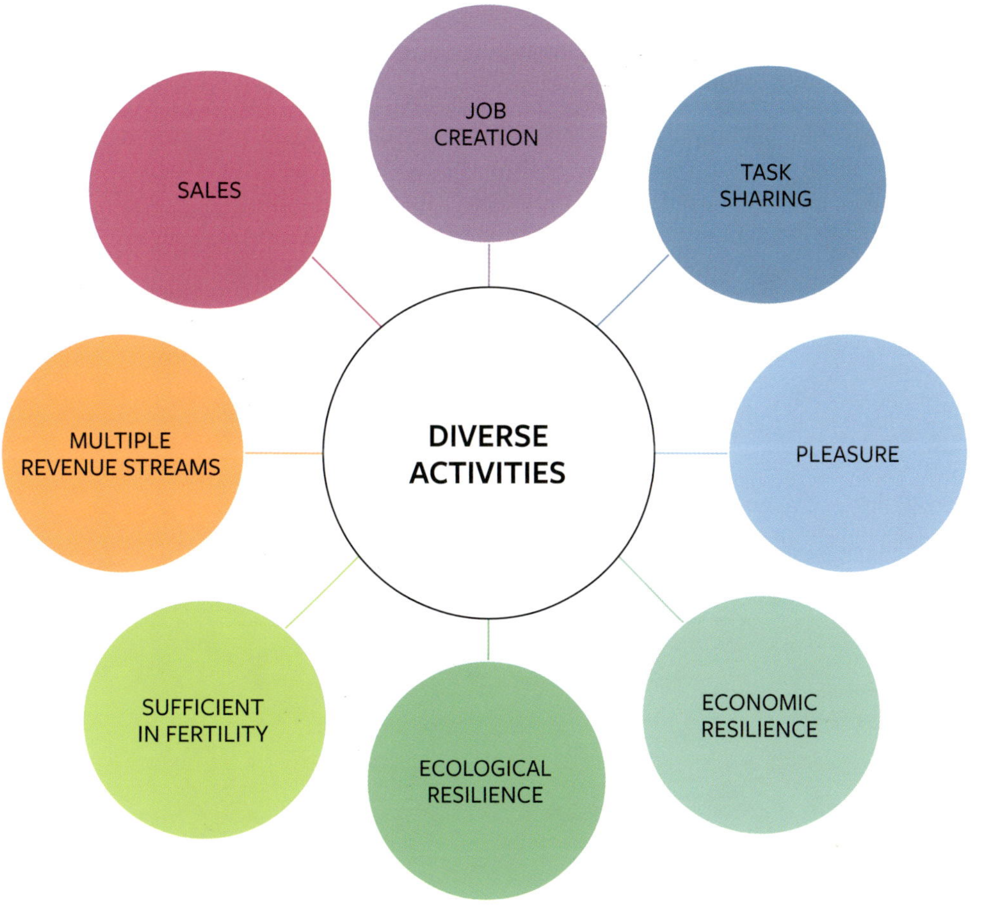

▸▸ *Brewing apple cider from a press that travels from farm to farm. Cider is the official drink of Normandy.*

Multiple incomes: instead of depending on one activity, microfarms can carry out multiple activities. They can create one or more workshops for processing their different products to add value on site.

Marketing: it is generally more attractive to present a diverse range of agricultural products, including processed products, rather than a single product. Local customers will be more motivated to go to a store if they can find a wide range of foods, including vegetables, fruits and berries, aromatic plants, eggs, cider and apple juice, honey, jams and syrups.

Job creation: a 1 hectare microfarm can have several activities. A microfarm is a great source of local employment.

Sharing tasks: from the moment when two or three people work full time at a microfarm, collective work becomes possible, as does a sharing of activities. Vacations can be taken in rotation. Tasks can be distributed according to the tastes and abilities of each person. One person can be designated to carry out the management tasks or public relations for the farm. Team members are less likely to suffer from the loneliness that sometimes weighs on farmers. Overall efficiency of the farm will improve; the whole is more than the sum of the parts.

▾ *Millions of gardens in cities and suburbs could be transformed quickly into microfarms, without much investment beyond labour.*

Pleasure: for most of us, it is more satisfying to carry out a variety of activities, on a diverse territory, rather than a single task that can become monotonous. Acquiring new skills and mastering various activities is a source of personal growth.

Economic resilience: not putting all your eggs in one basket, by diversifying the occupations carried out on the site, increases the capacity to face obstacles. If, in one year, production of one or two sectors is affected by unfavourable conditions, the others make it possible to get by. Some investments may be shared and thus more profitable: a vegetable stand, a computer, a website, a draught horse, to name a few.

Ecological resilience: microfarms are an oasis where multiple small environments work in synergy. Countless connections are established. Trees, hedges, and ponds establish a microclimate more favourable to animals and plants. The biodiversity of the site increases. All of these elements improve the capacity of the microfarm to resist climatic hazards: strong storms, drought or prolonged rain.

Sufficient in fertility: the more the site matures, the more biomass production increases. It becomes less necessary, or even no longer necessary, to import organic matter from outside. The microfarm becomes self-fertile and self-sufficient.

However, it is important to ensure that diversity does not become a factor of failure: each new activity requires investments, skills, time… Be careful not to spread yourself too thin! The main activity, the one that generates the most income, must be clearly identified and favoured.

On a plot of 1 hectare, it is possible to create several full-time activities. This constitutes a major opportunity, but one that should be managed wisely because it involves potential risk. We will return to this in the chapter devoted to the human engine of the farm (p.163).

◄ *Milking dairy cows in the fields is a practice that has almost disappeared. It is a surprising sight for these visitors. Living countryside, cultivated with respect for the Earth, is a major asset for the development of territories.*

16

THE FARM AND ITS ENVIRONMENT

We believe that to have the right to dominate the Earth is wrong-headed thinking. You don't believe that you have the right to dominate your father or mother. On the contrary, you have great responsibilities towards them and must treat them with love. Instead of persisting in cultivating the spirit of possession, we would do better to be constantly in search of harmony. This is what we must teach our children above all.

Dhyani Ywahoo, Etowa Cherokee[70]

Societal and Ecological Functions Fulfilled by a Microfarm

We have seen that a farm can provide immense services to the land in which it is located. However, the images that form in the mind of a person who hears 'farm' are: milk, meat, cereals, fruits and vegetables... Few people perceive all the functions fulfilled by an ecological farm! Hence the importance of highlighting these services, whether in a business plan or during exchanges with local elected officials, politicians and potential partners – including our customers.

Society should revisit its relationship with the land and those who cultivate it. Farmers have always been at the bottom of the social ladder. Even today, nearly 80% of the world's malnourished people are food producers![71] This shows how little they profit from their work. The major services rendered by farmers warrant more recognition and support from the community. Is it normal that setting up a farm is still such an obstacle course?

▼ *Agriculture is the foundation on which an economy is built. However, this sector is most often neglected, even despised in the policies of development. It results in social and environmental disasters. An important part of food produced in the world results from the work of peasants with bare hands, often shouldered by their out-of-school children.*

Knowing the functions performed by a microfarm can therefore fuel your argument, but also – and this is just as important – give you greater confidence in your enterprise. When we see the accelerated degradation of the biosphere, becoming a farmer can be seen as a vital mission. This is not a call to take yourself too seriously, but to give this your best shot. When difficult days arise, awareness of the value of the mission drives courage.

A Human Ecosystem on the Farm

The success of a farm depends on many factors, notably the capacity of farmers to fit harmoniously into their eco-friendly fabric, economic, social and cultural.

We can install the most beautiful agrarian system which, if it is not accompanied by the creation of a united and resilient human community, is not worth much in a crisis.

It is difficult to be totally invested in the creation of one's farm, and at the same time to devote energy to welcoming and exchanging! However, it is probably better to do a little less in the gardens and devote more time to human relationships. This means paying attention to every detail: the clarity of the website, the tone of the message on the answering machine, the flowers at the entrance to the farm, the order and cleanliness of the store, the smiles, the quality of the presence at the markets, memorising the names of customers...

When we undertake a project, many things are beyond our control. We can never control the weather, the macroeconomic situation, or the price of food... So, let's take care of what is in our hands. Nothing is insignificant. The love of the little things has a powerful leverage effect on the big ones.

Be Attentive to Every Aspect of the Project

Undertaking a slightly ambitious project is like building a cathedral. I can only imagine how the medieval builders must have felt, investing decades of hard work before seeing the completion of their work. If they wanted to go the distance, to work for a lifetime without losing motivation, building a cathedral required them to let go of the result: it was the process of building that was important in itself, the construction site that was a source of joy.

To focus too much on the final result, the time spent cutting a single stone would have been seen as senseless. Yet every stone counted. Not one could be missing from the building. The masons and carpenters built themselves up in an asceticism driven by the desire to give their best. The construction site led them to surpass themselves, to become better.

A farm, like a cathedral, rests on a number of pillars. It would not be very wise to put all one's energy into raising one pillar and to leave the others flush with the ground: the building would be completely wobbly, impossible to lay!

Better to go slower and give greater confidence in your enterprise. When you see the accelerating degradation of the biosphere, becoming a farmer can be seen as a real mission. When you gradually build up each pillar.

Each sector of your activity deserves a share of your attention. The crops, the planting of trees, the weeding, but also the communication, the relationships with your customers, the links to be woven with your community, the cleanliness of the store and the toilets... If you put 100% of your energy into agricultural production and your management is catastrophic, you are building an unbalanced edifice! Even though it is more difficult, cultivating a holistic vision is a guarantee of success.

Connections for Success

When you come from a non-agricultural background and move into a rural community, especially to start a farm, you can be met with a lot of suspicion. It must be recognised that people from the countryside have long been despised by people from the cities and that they have sometimes retained a mixture of frustration and shame. Even if society has evolved, this leaves traces.

It is important to understand that city dwellers; who come to the land with different practices to the local farmers, will often be mistrusted. And if they also claim that their agriculture is better, they become downright unbearable!

This calls for humility and discretion. Do not try to change your new community; do not give any lessons! Prove yourself patiently. Work hard, always be available to help, but don't overdo it. Over the years, recognition of your work will come and you may be asked to participate in the local community in some way. You will eventually discover that what you may have thought of as defiance may have been a lack of self-confidence and a profound kindness.

A farm makes its living by selling its produce, but the flow of money is only part of the story. Over the years, networks of exchanges and services, of gifts and counter-gifts, as the ethnologists say, are woven that contribute greatly to the economic and moral success of a farm.

Know how to give, know how to receive! As I write this, I am reminded of all the people in our community who have given, in various ways, sometimes a lot, to help us get through the difficult years. It is thanks to them that we are here today.

▲ *Small-scale, family-based agriculture sometimes perpetuates traditional practices that allow the enhancement of regions where environmental constraints are strong, such as mountainous areas. These activities are full of memories and often allow for quality production. The Appellations d'Origine Contrôlée (AOC) and other labels help them to survive.*

▼ *A growing number of new farmers, especially the younger ones, aspire to a collective venture. The creation of a farm with several people solves many problems, but it also poses other problems and requires the implementation of good collegial governance.*

◄ *Perrine talks with a chef while harvesting. Only small-scale structures are able to create strong ties of proximity and offer such a personalised service.*

17

THE 'HUMAN ENGINE' OF THE FARM

Our prayer is for a good and happy life, for an abundance of soft rains to obtain good and abundant harvests. We pray for the balance of the Earth, to live in peace and to leave a beautiful world for the children to come.

Dan Evemhema, Hopi

The crazier the project, the more seriously we have to do it.

We go as far as we can, but we don't go haphazardly anywhere or anyhow.

The success of a farm comes not so much from the capital invested as from the quality of its human engine. The person you are, with your dreams, your talents, your dedication, your integrity, is more important than the size of your portfolio.

Inner Vision

The main human engine of the farm is you. Beyond the training you may have received and the professional skills you may have acquired, your determination, your ability to take risks and to play the game to the end, whatever the difficulties, will make the difference. Cultivate this precious ability and don't lose sight of the vision that drives you in the midst of everyday worries. Your inner strength will be a source of leadership and will enable you to unite energies around your project.

Inner strength comes from a good alignment of your objective with your inner self: if it fits you perfectly, you will be able to deploy energy that will surprise you. But how do you know if the project is really your calling, if you are not entering into someone else's dream? The suggestions made in volume one will help you to see this clearly. And then there is listening to your little voice, your inner compass.

Most people focus too much on money when they think about creating a project, ignoring the value of ideas and dreams. Your ideas and dreams are energy, and the higher and more generous they are, the more powerful that energy is! It moves the universe around you and attracts responses, collaborators or partners, and even the necessary material means.

If you feel deep down that what you are doing is deeply right, ask for what you need, get to work and move forward. It will all come at the right time.

Polish your dream patiently, like a diamond. Spend as much time as you need to groom the idea. When it's time to give birth, take action and jump into the unknown without looking back! Most people, when they dream, don't allow themselves to dream big enough and, when they work, aren't rigorous enough. Be a pragmatic dreamer! Dream big and work hard. A farm is an object very much embodied in matter, you have to know how to manage this material dimension. In France, it takes twice as much capital to create a job in the agricultural sector as in the steel industry. In order for the dream to be embodied in the real world and to transform it, one must accept this burden, cultivate the ability to take care of the material, of money, without losing the vision.

You can never be too idealistic; you can never be too concrete. One cannot go without the other. On my sailing school ship, I followed two formulas that I often repeated to the crew:

"The crazier the project, the more seriously we have to do it."

"We go as far as we can, but we don't go haphazardly anywhere or anyhow."

If you are in your right place, well aligned with your dream, if it really corresponds to the person that you are deep inside, then there is no reason why you should not overcome all obstacles!

▼ *Creating Bec Hellouin Farm took us to the limit of our strengths. Especially in the early days, the example, advice and friendship of organic farming pioneers such as Pierre Rabhi, Philippe Desbrosses or Eliot Coleman helped us to continue.*

> "Choose the words you use carefully, they create the world around you."
>
> A Native American

The First Three Years: The Test of Truth

Starting a farm in today's environment is a very difficult adventure, especially in the first three years. You are vulnerable during this period when you are faced with many obstacles: you are not yet well trained, you have to start investing while the farm is not in full production, the soil is not yet soft and fertile, it is full of weed seeds, you have not yet built a good commercial network...

It is better to be well aware of all this before starting and to prepare yourself mentally and technically. Some people go through this period gracefully, others leave their dream, their savings, their marriage and throw in the towel.

There is no need to complain about the difficulties: you were warned before you started and you chose this project with full knowledge of the facts. We all have a tendency to complain, to blame the outside world (the weather, the price of fruit and vegetables, the government, the customers...). This weakens us and takes us away from the solutions. We have to take full responsibility for what happens to us and put all our energy into getting out of the mess! Make every difficulty an opportunity to make progress.

Your mistakes are precious; each one of them teaches you something, if you want to learn.

I have absolutely no affection for Napoleon, who caused the death of millions of human beings, but I love this quote attributed to him: "Victory belongs to the one who lasts a quarter of an hour longer." I have often observed, in my life as a sailor and then as a farmer, how true it is. I have seen many people give their all to a project that was particularly close to their hearts, and then end up giving up as the difficulties accumulated, when they had never been so close to the outcome.

▸ *While designing this manual, I am often worried about the fact that it presents mostly pictures taken during the nice season, when the weather is warm and everything is fine... When it's raining and freezing, we rarely think of taking the camera out! I hope this will not mislead readers by presenting an idyllic version of our job. Because being a farmer requires a good dose of moral and physical strength. I pay tribute to the exceptional courage and endurance of my partner farmer: she held on and went through all the crises and the hard times that marked our adventure.*

Awaken the warrior in you; don't fight **against** a system or **against** others, but **for** a better world, **for** future generations to live in a restored planet. You have no enemy. You cannot afford to divide your energy; invest it all in the pursuit of your goal!

Share the Vision

Learn to articulate your vision. Share it in clear and simple words. There is powerful energy in words.

Step out of the coldness of the impersonal; don't be afraid to divulge something about yourself. This can be articulated in your business plan, on a sign that welcomes your visitors, and by the pictures on your website. Choose the images you share. Don't be naive; your farm sends a multitude of signals to its surroundings and visitors. A pile of old tarpaulins lying around at the entrance of the farm or, on the other hand, flowers and a nice sign to the parking lot are all signals. So choose wisely!

If you are sincere and honest, you will touch what is most sincere and honest in the heart of your interlocutors and engage a relationship on these values. You will be surprised to find that clients, trainees and employees who share the same values will naturally come to you. This avoids many misunderstandings.

I notice that at the farm, our team is made up of very diverse personalities, but they have common traits: humbleness, great sensitivity, kindness, integrity, total investment in the work and benevolence. Our team co-opts each other and passes on its values to newcomers, without the need for any formalisation. As with any alliance, there are sometimes tensions or bumps in the road, but in the end everyone wants to work towards a common goal. The day visitors often tell us how touched they are by the quality of the welcome, and the trainees who come for training testify to the depth of the human experience. It is not surprising, then, that the farm is an excellent marriage agency! Over the years, we have had the joy of seeing many couples form and observing the pretty faces of 'Bec Hellouin' babies!

A Job Well Done: A Solid Foundation

What unites a good team is a well-defined project and the pursuit of a job well done. Laying the foundation is hard work. Build into your farm a strong requirement of professionalism. A principle of realism. The farm is not an ethereal community but a working tool that makes it possible to generate money to pay salaries at the end of the month. We must know that we can count on the other members if they say "I will take care of the watering" or "I will feed the animals", that they will do it. Day after day, respecting commitments makes it possible to build something more: of respect, of esteem, confidence, complicity, and finally, friendship. This is how, in partnership, we can achieve the vision that drives us. Not the other way around: groups of people who operate on an ideal of friendship may neglect rigorous and regular commitment on a daily basis.

▲ *In all honesty, working as a couple was very difficult for us: two strong characters in the same boat can cause sparks to fly! Working together adds additional challenges to married life, which is not easy in itself – I suppose you will agree. But the shared trials constantly push us to know each other better, with trust and respect.*

Working Alone

Many market gardeners set up alone, initially anyway. This requires total investment during the start-up phase which can last three to five years. The cost of labour is obviously reduced and the tax is lower, as long as remuneration is non-existent or low.

One of the advantages of working alone is that it allows you to become familiar with all the tasks to be carried out. We are not necessarily good at everything, but at least we know how to do everything. This gives legitimacy and an aptitude for training others when building a team. However, starting alone requires huge stamina.

We often see farmers who do their agricultural activity alone while their spouse maintains a different profession for financial security.

Working Part Time

The small size of a microfarm means that it can be worked part time, but this makes it more difficult to obtain the status of farmer. Part-time farming has a number of advantages. If you love your current job and earn a decent living, the part-time option allows you to be on the farm less, without the pressure of its economic success. This can result in a good quality of life. We can imagine doctors, teachers, traders, employees who are also part-time farmers. Self-employed professionals are particularly suited to this: artist, director, consultant, writer, psychologist and more.

Working as a Couple

Setting up as a couple allows you to better support each other and share the difficulties. However, it is necessary to have previously amassed sufficient savings to support a family, while attempting to make the farm profitable.

Regulations allow one person to be the farm manager while the other holds the status of collaborating spouse. This allows you to benefit from minimum social security coverage in exchange for low contributions, but the pension is almost non-existent. For example, it is worth creating a 'GAEC' (agricultural group joint exploitation) between spouses, each being operations manager for a workshop (one in market gardening, the other in chickens), in order to guarantee better benefits and real recognition of the work of both members of the couple. Research the financial possibilities in your country – they vary but there are always opportunities to maximise yields. It is obviously necessary to ensure that fatigue and professional stress do not hinder good relations of a couple who live and work long term all year on the same site. Daily vigilance is necessary: the couple's relationship is the priority, not the work!

Working in Groups

When we create a farm, we see that the workload grows quickly, especially in a diverse system. Working alone or as a couple, we might want to share this load. Building a team can take several forms that can coexist: hiring employees or associates, or welcoming trainees.

Training Interns, WWOOFers...

An organic farmer can very easily accommodate trainees in professional training or interns, like WWOOFers for example. The requests for these positions are numerous.

This welcome must be based on clear and healthy agreements. Too many organic farmers think that volunteer interns can solve a problem of work overload and economic insecurity. This may be true occasionally, for a major project, such as planting an orchard. However, from experience, we think it is problematic to ask interns to make up for a shortage of manpower. This is an unsatisfactory solution. If the workload is too high, better to review farm design than rely on inexperienced labour that is there to learn, not to make up for an employee that you cannot afford. When we welcome an intern, there is a contract between us. It's a reciprocal relationship - interns come to learn, you owe them time, availability, and the tasks you entrust to them must be varied (too many trainees are confined to weeding!). In exchange, they offer labour.

Be aware that each trainee will take up some of your time and you will have to adapt. If you regularly welcome new volunteers, you repeat the same instructions week after week, teach the same techniques... These interns will always work slower than you and they will make more mistakes. Eventually, many farmers stop welcoming interns when they are fatigued and weary from a constant turnover of interns.

When you supervise more than three or four beginners, it absorbs almost all of your time and you are no longer performing at your usual level. At the end of the day, it's a safe bet that they will not have progressed further than you would have done alone, and you will be more tired! Let's be clear: we are opening our farm to the trainees out of a desire for welcoming and sharing, to experience great encounters, to train the farmers of tomorrow. They offer us their enthusiasm, break the solitude and alleviate the monotony of certain tasks. Beginner trainees do not replace a good salaried employee, but it is nice to work in pairs or threes in the garden, sharing a passion for this profession and discovering their life experiences. Farmers don't often leave the farm and these young visitors are windows to the world. We received a lot of interns at Bec Hellouin and, paradoxically, we have often been criticised for this, as if the volunteers kept the farm running and explained its results! But how do we usher in a new generation of ecocultivators if we barricade ourselves behind our hedgerow?

The first years we gave everything, brought WWOOFers into the home of our children and received them at our table seven days out of seven. We hosted some extraordinary people, but also unbalanced folks, or alcoholics. Quite frankly, this disrupted our family life. We learned to manage hosting and no longer encounter these problems. Firstly, we decided to host interns for a minimum of one month. We requested a cover letter and a CV in order to study the applications. We also clearly announced what to expect: market gardening is a difficult profession and demanding, and we expected our trainees to commit to a real investment. This has reduced turnover, discouraged dilettantes and allowed us to choose the most suitable candidates.

Secondly, we stopped hosting WWOOFers and limited the invitation to future farmers engaged in professional training. We do not offer overnights on the farm. Those who come do so because they really want to learn. Currently, we are no longer accepting interns outside of training periods, when the farm is bursting at the seams, so that we have breaks, times when the farm is calm. This rejuvenates us to be fully available during training. It's like in music: you need silence to appreciate the notes. Sometimes we make an exception for one or two people for a long internship on the farm, usually for one year. These people are totally integrated into the team and they really want to learn the job.

We have often noticed that being demanding does not slow us down, quite the contrary; we avoid all those who want to treat themselves to a cheap farm holiday. Only the most eager to invest and progress are invited. We suggest establishing rules. In all cases, care must be taken to have public liability insurance and respect existing legislation, in particular health and safety rules. In France, legislation strongly limits the reception of volunteers on a farm or a business; volunteering is normally reserved for non-profits. It is imperative that having trainees is ruled by an agreement of internship or WWOOFing.

▼ *Nowadays it is difficult to connect with a farm, but new installations are often accompanied by a desire for openness. We have been very welcoming and beautiful encounters and friendships have endured. However, we have reduced hosting outside of training because it requires a lot of energy – energy that sometimes we just don't have, with the fatigue accumulated over the years.*

◄ *Fabien, our chef, speaks with Teddy, the crop manager for the past two years who is also from the culinary world. The presence of a stable and competent salaried team has made it possible to face the multiplicity of tasks and requests. We have a lot of gratitude for our employees and our collaborators. However, it is necessary to create enough value to pay everyone a remuneration corresponding with their commitment. This is a challenge in the current context.*

Hiring Employees

In the event that the farm is managed by a couple and it becomes necessary for another person to work on the site, new hires should have the status of employee. Many farmers need additional labour, even if only seasonally. However, hiring employees during the initial stages of the farm must be done with a lot of discernment: put your foot down - in France the cost of labour, with social charges, is such that you will have a hard time paying for them with the sale of your products. Organic fruits and vegetables are not very profitable. It hardly allows the farmer to pay decent wages, to retain well-trained and competent employees.

If you only pay the basic salary, you attract people who are complete beginners in market gardening, or people aspiring to create their own farm, who will come to your farm for a period from one to three years with the aim of training, then set out on their own. It is a form of saving, but in these conditions, it is difficult to build a stable team.

If you pay a decent salary, comparable to other professions, you attract highly invested collaborators and dynamic team players, capable of taking initiatives and working autonomously. The sale of vegetables is not enough unless it is for a team leader on a high yield farm.

WWOOFing[72]

WWOOFing (World Wide Opportunities on Organic Farms) is a concept of eco-volunteering. Individuals spend time on organic farms to discover and experiment with alternative lifestyles, exchange, share, occasionally participate in certain tasks, learn techniques, acquire knowledge and know-how, and contribute to the development of biological, biodynamic and permaculture movements. The experience of WWOOFing is above all a life experience, a vector for social bonding, based on an exchange of best practices: the WWOOFer is offered room and board in exchange for their one-off participation in activities on the farm. Under no circumstances should any hint of subordination or hierarchical relationship between the WWOOFer and their host be established; the WWOOFer is not bound by any monetary obligation. Otherwise, the host farmer could be prosecuted for concealed work.

A presumed non-salaried activity, in the same way as volunteering, WWOOFing has no legal status in France. It is simply accepted. The WWOOFer is not employed or remunerated, whether in kind or in cash. They do not benefit from any protection.

For farmers who welcome WWOOFers, an insurance policy is strongly recommended to cover any damage they may be liable for. Furthermore, it is also recommended that farmers request civil liability insurance from the WWOOFer. It will be in their interest to have good coverage, for illnesses, accidents and their consequences.

Some colleagues say that organic market gardeners are exploited in France. The average remuneration in the profession is 800 euros net for 50 hours per week on average, stretched over the year.[73] It can be difficult to generate enough to pay a salary and cover expenses while we struggle to pay ourselves!

At Bec Hellouin, we are lucky to have two full-time employees. One of them has been with us for more than 10 years. They are responsible for the maintenance of the farm, which they know like the back of their hand, handle the market gardening, the animals, and collect data within the framework of our research programmes. They provide valuable stability.

We have had three successive heads of planting, younger employees who trained with us and who stayed for two to three years before leaving to create their own farms. This was defined by a common agreement before their employment.

Now, Perrine and I have taken over responsibility for market gardening with our full time employees and our chef, who has been full-time for more than five years and demonstrates the flexibility and versatility to justify his position. He also handles the secretarial and management duties.

It takes time to get used to the particular context of a farm with active and very diverse lives. Versatility, flexibility, investment, spirit of initiative and perseverance are all essential qualities.

In view of the above, try to delay the moment when you will have to create a new employee position, even if it means using service providers. If you start modestly, avoid hiring because all that you earn will go to paying salary.

▼ *There is an obvious alliance between a vegetable garden and the kitchen. The presence on the farm of a full-time chef allows us to explore this link. Hiring Fabien became possible the day our training reached full capacity. We never regretted it!*

An activity that reaches its cruising speed and leaves a comfortable margin risks becoming a deficit if you hire someone.

But there are also of course advantages to hiring. Governance is relatively simple with one or more employees and the departure of one of them does not jeopardise the sustainability of the farm. If you start with enough capital, help with certain tasks allows you to concentrate on more pressing needs.

Create a Farm on your Land without Working on it Yourself

Some decision-makers and entrepreneurs become aware of environmental problems and decide to take action. We are often contacted by owners of land and financial means, who aspire to create a relaxing microfarm on their estate, without working there themselves. This can be a risky venture if the owners do not have the skills or knowledge to supervise their farm. If they hire a person who is very well trained and autonomous, they may have trouble retaining that person and meeting the level of salary requested. If they hire untrained people, the farm will have operational difficulties.

In addition, professional market gardeners, well trained in permaculture market gardening, and available for hire, are very rare and coveted! Most often, they don't want to earn a salary and prefer to create their own farm.

Land owners may also wish to make their plot available to a market gardener. Here again, we encounter difficulties: if the land is rented with an agricultural lease, the farmer is safe and can invest, but in France rural leases are such that it will be difficult for the owner to terminate them in the event of misunderstanding.

The provision of land on a precarious basis protects the owner, but the farmer will not benefit from the investment made to create a microfarm: biotope development, planting of trees and more.

This is an obstacle to many potential projects and it would be useful if legislation would evolve and other options were possible. These questions are complex. Better to be realistic and consider this type of microfarm arrangement as a form of patronage, investment in the long term, for the benefit of the farmer's family and the local community, or as a means of adding value to a property, rather than as a profitable but temporary business by itself.

◂ *I can't see this photo without feeling a kinship with each of these young people who sparkle with intelligence and humour, so courageous in their investment, so rigorous in their work. Many of them met their true love at the farm, an additional joy!*

Collective Set Up

In France, many project leaders prefer to partner with several people in order to share the work and have a better quality of life. There are suitable legal arrangements in France, such as the GAEC (collective farming), that permits an association. It's important to find the best form of governance for the group and ensure that the departure of a partner, even if only in retirement, does not weaken the collective.

In the small world of permaculture, many associations are more or less informal, not governed by a legal statute. Many projects fail due to the human factor. Everyone's expectations must be fully made aware and documented. A collective project must be drawn up, as well as a common charter (refer to volume one). Here are some points to keep in mind when drawing up a collective farm plan.

Capital: if the microfarm is created by several people, who provides the necessary funds? Questions of money are a factor of potential dissent and it is important that the foundations of the agreement are transparent, healthy and solid. All vague parts of an arrangement risk jeopardising the sustainability of the arrangement. We cannot recommend enough that all points are recorded in the form of contracts, and written with a legal advisor if necessary. This avoids misunderstandings later.

Distribution of income: in the case of a team comprising one or more employees, things are clear: the employee receives what is due to him contractually and the farmer is remunerated with what is left at the end of the year, after the deduction of all charges.

In the case where several partners work together, the question of income distribution must be clearly defined, as early as possible. Divergent views may arise and become a conflict: how much should be paid to each partner? What share for investments? So-and-so works less than the others, should he earn as much?

It seems that in Europe as in the Anglo-Saxon world, associations, communities and sustainable ecovillages have gradually become subject to governance tools, charters, management tools and regulations similar to those of businesses. Let's hope that this is not in contradiction with the initial vision! Such tools become what we do with them: they can serve the cause or stifle it; it's all a matter of discernment.

Being a farmer today is anything but a solitary adventure! We come out of a model that isolated farmers and enter a new era made of work and shared dreams.

Design and Redesign!

A good design, reworked as often as necessary, allows you to avoid finding yourself trapped in a situation that does not suit you. Too much diversification can create extra work or chronic overload that keeps project leaders from achieving their initial goals. The risk of finding yourself leading a team or stuck behind a computer is a real possibility. It requires a certain amount of lucidity and courage to straighten out a situation that is getting out of hand and reposition yourself. Sailors know well that no ocean crossing is made in a straight line: according to constantly shifting winds, we constantly readjust the trajectory.

Ferme Biologique
Bec Hellouin

◄ *Fabien will never say it himself because he is modest, but the meals he prepares from our products are Michelin-star restaurant quality. Visitors often ask him to write a cookbook. We are proud of the way he values the work of our team. The fight for organic food is won through simple, but extremely tasty products.*

18

PREPARED FOODS

The question of food security in cities will require a profound modification to rural zoning. We know that the transition to a post-carbon world will either be planned or forced. But, in both cases, it will most likely be rapid and radical.

Pablo Servigne[74]

Diversify and Enhance Production

A microfarm can hardly afford to sell its production at wholesale markets. Selling directly and creating a relationship with customers is more in tune with the spirit of microfarming. Diversifying production and transforming some of it into other products ties in with a direct sales approach. This allows you to create more value within a small area. It is also a pleasurable job: there is something fascinating about transforming the products from your garden into delicious products. Knowing how to do everything yourself, from soil preparation through to processing, labelling and sale, is a source of personal fulfilment.

However, we must remain realistic and not try to do too much, too quickly. This is an error that we made. During our first season, in 2007, we produced vegetables, fruits and berries, but also jams, syrups, aromatic plants and dried medicinal herbs, cider, sauce, apple juice, cider vinegar and bread. I remember Perrine working until midnight to make the jams. We did not weigh up the constraints associated with this amount of production. This chapter would have been useful advice for us at the time and it would have allowed us to better evaluate the efforts required to accomplish all of this.

Skills

Each type of product transformation requires specific training. Some are easier than others: if making jam is within the reach of everyone, producing cider requires a certain skill set. You can teach yourself, by studying and trying, but a brief course or an internship with a professional can avoid a lot of trouble. Common sense dictates that we only sell products that we know how to make. We feel bad when a customer brings back mouldy jam.

Legislation

Food processing is regulated to guarantee consumers the hygiene and sanitary quality of food products placed on the market. When you prepare food as a profession, it is important to be perfectly aware of the regulations of your country. In France, we are subject to yearly inspection that examines our documentation, in particular recipe books and the batch number register. We are equally carefully inspected from time to time by the DGCCRF (Directorate General for Competition, Consumer Affairs and Fraud Repression).

▲ *We have a 100% organic diet and prefer to eat vegetarian. Just-picked vegetables keep the integrity of their flavour when eaten raw.*

Legislation Regarding Production

Each type of processed product is subject to regulation. The certifying agricultural bodies, professional organisations and the DGCCRF can explain these regulations.[75]

Some products have little or no risks for consumer health: products manufactured solely from plants, and in particular those sterilised by high heat, like confitures and breads. But every time an animal product is present, the risks increase and regulation becomes much more strict.

Recipes must be recorded in a manufacturing log, including batch numbers to ensure traceability.

Label Legislation

Labels are controlled by general regulations as well as organic food rules. A food product label must be clear to the consumer, readable, with specific important information: in general, the name of the product, its exact composition, the expiration date, the name and address of the producer and the batch number. Organic producers must display the organic logo, the number of the certifying body. When several ingredients are included in the recipe, the organic ones must be indicated.

Organic producers are, moreover, required to get approval for each recipe and label from their certification body. The controller will explain all of these steps during the first certification visit.

Infrastructure

A Location Dedicated to Product Transformation

Within a microfarm, we generally do simple and artisanal products that do not require sophisticated equipment. However, if the production is on a regular basis, the family kitchen is not suitable and we quickly realise that it is more practical to have a dedicated room, hygienic and easy to clean, equipped with a water source, where equipment and ingredients can be stored and kept in good condition. Fully tiled with stainless steel work surfaces is ideal. An adjoining room that serves as a storage place for containers and labels is also needed. Jam jars and bottles can be bought by the pallet to limit production costs, but they will need storage space as well.

If we process animal products (dairy, meat, eggs, etc.), the commercial kitchen must comply with more restrictive specifications. It's better to find out before construction, and seek the approval of veterinary services.[76]

Create a Production Space: Materials and Equipment

The workshop and equipment depend on production considered. Given the multiplicity of possibilities, we refer you to the databases available.[77]

If processing a lot, a labelling machine can save time. In this case, the labels must be printed in the form of rolls adapted to the machine.

▼ *A farm chef can transform production surpluses, so nothing is lost. Fabien's canning jars are not marketed because we do not have the necessary approval. They are consumed on the farm.*

The Economics of Transformed Products

When you start to make confitures in the evening, after a day in the garden, preparation cost is less of an issue. If the farm does it professionally, however, it becomes desirable, if not essential, to have a clear idea of the economics of each recipe.

If investment in specialised equipment and infrastructure is low, and the products are processed on the farm, they can bring in additional income.

But if we invest in a dedicated kitchen and hire labour, the net margin falls precipitously. We are then forced to produce significant volume to make fixed production costs profitable. But will we be able to sell this volume? If we have to distribute some via resellers, then the sale price is reduced by their margin (20-30% at least) and it becomes much more difficult to make a profit.

Better to avoid finding yourself straddling a small, diversified artisanal production and specialised production. It is generally not comfortable being halfway between two scenarios.

To avoid increasing costs, look for quality containers and labels, but at an affordable price. For the labels we quickly gave up trying to make them with a home printer. Instead, we designed them with Philippe Laborde, who also designed the French version of this manual. The templates are sent via the internet to online printers for small runs at a lower cost, like 100 self-adhesive labels.

At Bec Hellouin Farm, our chef Fabien makes a lovely range of prepared foods. He precisely calculates the cost of each recipe. The profit margin is very low, a few pence per pot. However, Fabien is hired year-round and the products created allow him to make the most of his time between two training courses. Raw materials are generally surpluses that would be lost if they were not processed. Finally, the offer of processed products makes the store more attractive, especially in winter. For all these reasons, processed production is justified, even if the margin is minimal.

Setting the selling price of a processed product can spark lively debates! The sale price results from an arbitration between cost and margin and the 'psychological price' that customers are willing to pay. This depends on the packaging, the reputation of the product and farm, and its quality.

Organic market gardeners are often embarrassed to sell their production at a decent price; they aspire to sell 'as cheaply as possible to put their products within everyone's reach'. Is this fair? Considering the cost of the ingredients, packaging, labels and all charges that the farm must pay, a sale too low often does not provide a good return. Is it normal to live with the anxiety of bank overdraft? To work year-round without a vacation with your family? The price must be fair for the customer, but also for the producer. At a time when we are spending as much money on our mobile phones as our food budget, selling our products at a fair price will not make our customers go broke.

Recipe Development

When developing recipes, highlight the things that make a microfarm so attractive: organic, the region, and the artisanal side of your production. For example, we make a calvados cider jelly that is popular for its reference to the gastronomy of Normandy.

We are sometimes forced to use ingredients that are not produced on the farm, sugar for example. Organic farming regulations in France stipulate that at least 95% of the ingredients, excluding salt, water and additives, must be certified organic. We will now give some points of reference for the most commonly processed products made on a family farm.

Jams, Jellies and Compotes

These preparations use sugar and heat as a means of preservation. Sugar content is regulated to meet the requirements for confiture.[78]

To make jams, jellies and compotes in small quantities, no specialised equipment is needed and home cooking may be enough. If we launch into a batch of several hundred jars, it is best to have a seed remover and a jam cooker which stirs the preparation continuously.

We can improve the preservation of preparations by reheating the full sealed jars for a few moments at high temperature, in a steriliser.

Fruit Juice

It is possible to sell fruit juices, small fresh fruit or vegetables at the farm stand, on demand, in the form of smoothies for example. But to be able to keep fruit juices for more than a few hours, they need to be pasteurised. This requires equipment that is not within the reach of a small farm: a press, a system

▲ *Fabien also makes products that do not require sterilisation and they are sold in our farm store: jams, compotes, chutneys, and more. He creates original recipes that promote seasonal products.*

▲ *Apples, abundant on the farm, are transformed into juice, cider and vinegar. Offering many varieties of apple juice increases the attractiveness of the product line.*

for heating the juice and a bottling line. Better to use a specialised service provider. We deliver fruits and we collect full bottles with an expiration date stamped on them. They can even be labelled and put in boxes there.

You can make delicious apple juice with cider apples. To make our juices more attractive and take on different flavours, we make single variety batches. When you squeeze apples or other fruits, it is important to be vigilant about conservation. The apples are kept well a few weeks after harvest, however some get damaged and must be removed because they can transmit a mycotoxin called patulin. For fruit juice produced from apples, the standard legal limit is 50 micrograms of patulin per kilo of juice. All it takes is a few damaged apples per tonne of fruit to surpass this threshold. Contaminated batches must be destroyed.[79]

Cider

Making cider is a simple, natural process of transforming apple juice. But managing to obtain a pleasant flavour with just the right amount of fizz takes practice. To make a good cider, you need different varieties of apples, both tangy and sweet. We crush apples as late as possible in the season because cold slows down fermentation and the flavours become richer.

The first step is to wash, and remove any damaged apples. Then crush them because apples are too hard to be squeezed directly. The artisanal method uses a screw press. The taste is better than with a hydraulic press because less juice is extracted and the slower pressing delivers a deeper coloured liquid. However, this operation requires several hours of work for a limited quantity of juice. Farms that produce 2,000 to 3,000 litres of juice, like us, hire a travelling press.

The juice is put into carefully cleaned vats and valves. Any leftover remnants will transform the entire batch into vinegar. About forty-eight hours after pressing, pulp residues and other impurities begin to float to the surface and form what we call 'the brown hat'. To remove it, the juice is drained from the bottom and transferred to another tank. This gets rid of impurities.

Apple juice contains approximately 50 grams of sugar per litre. Natural yeasts in the juice will eat this sugar and transform it into alcohol. The process lasts from a few days to a few months. If it goes too fast some of the yeast is extracted. It can also be filtered. Cider is frequently weighed to monitor its fermenting. Bottle according to sugar content; sweet cider contains more than dry cider, and is less alcoholic. The most delicate operation is bottling and obtaining the right flavour. Yeast continues to be active for a few weeks when bottled. It releases CO_2 gas that makes the fizz. Too much yeast will cause cider to be overly carbonated and explode in the bottles or cause corks to pop. If yeast amounts are too little cider will be flat. There are different processes to control the amount of yeast. The most effective is counting particles under a microscope. If there is too much, the cider is filtered. When not enough, wait to let them multiply.

Making cider is exciting and rewarding and adds value to the produce. However, it requires having a cellar, vats, a filtration system, and more. It is done in winter, the only season that market gardeners can slow down a little.

Pear (perry) cider is obtained from particular varieties and is more difficult to make than apple cider due to the bitterness of pear juice. Pommeau and calvados are aperitif and digestive drinks made from cider. They have a higher alcohol content and that comes with regulations.

While Normandy is the country of cider, other regions have a beer or wine culture. Micro-breweries are multiplying in France. We do not have experience with these products. Perrine is hoping that a microbrewery

► *Our farm shop offers simple and natural products: no frills, our customers are looking for an authentic experience.*

will open near us so that she can recover the waste that is rich in microorganisms and yeasts.

Cider Vinegar

Making apple cider vinegar is child's play: just put mother yeast, a culture of acetic acid that forms naturally in cider, in a vat of juice. It looks like white plastic skin. The mother will expand until it covers the juice completely. In less than a year cider will transform into vinegar without any intervention.

The term 'vinegar' can only be used for liquid containing a minimum 5 degrees of acetic acid.[80] Laboratory analysis determines the amount. For small farm production levels, the cost of these analyses would be prohibitive. We can simply call it 'cider vinegar' instead of cider. Cider vinegar sells for more and can be flavoured with elderflower, red fruits and more. Moreover, both for our health and for farm needs, cider vinegar is a panacea: it is used to de-worm animals and as a seedling sanitiser. Don't place vinegar in the same cellar as cider due to risk of contamination.

Dried Fruits and Vegetables

Certain fruits and vegetables lend themselves very well to drying, a method of conservation that preserves nutrients and intensifies flavours. Various drying processes are described in volume two. In warmer countries, producers can use a solar dryer. We use the residual heat of the bread oven to dry apples and cherry tomatoes. They are then vacuum-packed into plastic bags to improve shelf life. Tomatoes can be jarred in oil.

Aromatic Plants and Dried Medicinals

Aromatic and medicinal plants can be dried, put into bags and sold. This is simple to do and does not require sophisticated equipment.

Bread

In France, commercial bread making is reserved for bakers who have artisan credentials. It is also possible to be a farmer-baker if you produce at least 70% of the flour used (see Chapter 9, p.93).

Preserves

Preserves are subject to strict regulations because there are health risks. It is imperative to use an approved professional autoclave (steriliser) that is capable of raising the temperature above 100°C – canning is done at 121°C. The equipment for this operation is very expensive and complex to use, so few small farms have it. Small batch producers should hire a service provider or use a shared kitchen.

Beyond its commercial appeal, the preservation of farm products helps to satisfy our quest for autonomy, and we are proud to serve a dish that was made entirely with our own produce.

▲ *Fabien's cider is appreciated both by famous chefs and our local customers.*

◄ *Located a few steps from the gardens, the farm kitchen is well equipped to prepare meals for up to 60 people. Between two training sessions, it is used to process prepared products.*

◄ *The farm shop was built in 2007 by a craftsman from the neighbourhood. The facade is made of stones and reclaimed wood; the rest of the building is built with wood from our forest. The cider cellar pictured is located under the farm shop.*

19

SELLING YOUR PRODUCTS

When it comes to the price of food, the word 'honest' pertains to both 'affordable' and 'fair'. For a small market gardener producing very high-quality vegetables, the challenge is to set the prices at a level for the farmer to earn a decent living – without being accused of dishonesty or elitism.

Eliot Coleman[81]

During training we emphasise two questions: "How many hours per week do you want to work?" and "How much do you want to earn?" A typical response: "Me? I don't want to sell! I only want to barter, exchange, share." Once, one of our students made a good point, saying, "Guys, we're missing the point here. We're not going to be able to use lettuce to pay for fuel or phone bills."

Beginner microfarmers typically want to move away from materialistic values that dominate our commercial society. Logically, we have to sell our products to get money. It's necessary for the survival of the farm.

▼ *The farm shop is also the dining hall for our farm team.*

Selling the Harvest

Producing and selling are a must with professional farming. For a market gardener, the time spent selling vegetables represents 15-30% of total work time. There are many ways to sell. It is beneficial to think this through before planting your first seed.

Sales Objectives

The first step is to think about a mode of marketing that matches your aspirations and talents. Some producers like markets; they flourish in a busy market and enjoy the customers. For others, the farmer's market would be an unbearable constraint; they have no interest in direct sales and prefer to put their work into production, even if it means selling out at a less attractive price, from resellers or on a wholesale market.

Market Research

Get to know the market in your area. How many producers are there? Where are they located? What products do they sell and how? How is the market organised? Is the market saturated? Are there producer organisations? Are certain products in demand?

Identify, Quantify and Qualify Local Demand

A market study should allow you to identify opportunities. Knowing which products are sought, by whom and in what quantities, will save you a lot of trouble later. It is much more comfortable to build your sales network before entering into production. It's not when the gardens are overflowing that we search for customers, a mistake that we once made.

Niche Strategies

It is eminently desirable to avoid entering into competition with established organic producers. A region may already have many market gardeners and the local markets may be saturated. Better to know and research what types of products are in demand in this situation.

Build a Realistic, Adapted and Evolving Sales Strategy

Geographic positioning greatly influences sales. Our farm is isolated in the countryside in a region with very little focus on organic, forcing us to send part of our production to Rouen and Paris. We sell to several different wholesalers, organic shops, restaurants and participate in a few community supported agriculture programmes. We concluded that no partnership is completely reliable and durable. It was necessary to constantly adapt to a changing market. It is best not to rely on a single client. We experienced this when the owner of several restaurants that we supplied went bankrupt.

Commercialisation

Sales Legislation

In France, direct selling is considered an extension of agricultural activity for farm-grown products and does not require registration. Farmers can resell products up to 30% of total sales. Above 30% requires a status of seller or craftsman.

It's important to know the regulations of your country and region. Whether you sell on the farm or at a market, there can be varying regulations. In France, only four products are regulated: poultry, cheese, eggs and milk.

Hygiene

Foodstuffs and animal products will also have their own regulations, and will vary from country to country. All entities in the food chain are subject to rules regarding the quality of marketed foods.

As specified in a French guide intended for organic market gardeners, "generally speaking, organic products offered on the market must be healthy and safe for consumer health. They must be clean, ripe and free of pesticides or other substances. There is a guide to good hygiene practices available at www.ladocumentationfrancaise.fr. Pay particular attention to cut vegetables, such as pumpkin that will be more likely to become contaminated than uncut produce. There is no specific guide regarding storage temperature, it must be adapted to products."[82]

Taxation

In France, profits from the sale of production are agricultural profits that enter into the category of social contributions even though the farmer uses commercial methods including commercial premises and staff. Profits made are not considered to be industrial profits or commercial activities. In cases where the farmer buys farm products for resale, sells for other producers or operates through consignment sales, the rule of industrial profits and commercial activity applies. Of course, the legal form of the company and/or the status of the farmer have an influence on the determination of taxation. A professional accountant who specialises in farm accounting will be able to advise you.

Labelling

Labelling is mandatory when selling on the market or on the farm and when production is sold to a shop or restaurant. For direct and retail sales, three mentions are mandatory. The labelling must indicate:

- The nature of the product;
- Its origin (at least the country);
- Its price (by weight or by piece).

As mentioned for processed products, labels must also bear the European AB and the approval number of the organisation certified.[83]

► *We chose a brightly coloured, classic look. The design was inspired by the Diderot and d'Alembert encyclopaedia from the 18th century. However, this image did not support production very well.*

Set your Prices

Setting your prices requires a price study of usual sales in your region. Record prices in markets and shops. In Normandy, the regional group of organic farmers, Giques, periodically publishes a market price (price list) for minimums, maximums and average prices offered by market gardeners in the region.[84] For certain uncommon products, like mini vegetables and edible flowers, research needs to be carried out among specialists. We happened to photograph the price list that restaurateurs access. We had several price scales depending on whether the products were marketed for direct sales or to restaurateurs. Now we have aligned all our prices. Wholesalers receive a discount. Dialogue with customers, resellers and restaurants helps define what seems fair to each party.

Make a Market Price List

For several years we have offered a weekly market price list of all products available for the week. It's quite a big job to prepare, but gives clients a constant update on what is available from our gardens. The list is a spreadsheet, with boxes for customers to indicate their choices. This way, the lists become purchase orders that are printed, for farm worker use when harvesting. Sometimes there are minor differences between what is ordered and what is delivered. The price list is then adjusted to reflect any difference and it also serves as a delivery receipt and resource to prepare the invoice.

Create a Point of Sale

Nowadays, there are few opportunities to visit a farm. Customers often appreciate shopping at our farm store, if it is not too far from their home. The shop is quickly becoming a friendly gathering place where loyal customers exchange the latest news. Our store is open once per week. Day visitors can take the opportunity to discover the farm, for an entrance fee.

Farm sales are probably the least demanding for the producer.

Some producers use a common point of sale. This approach has many advantages: a greater diversity of products, expenses distributed among several producers and customers are shared.

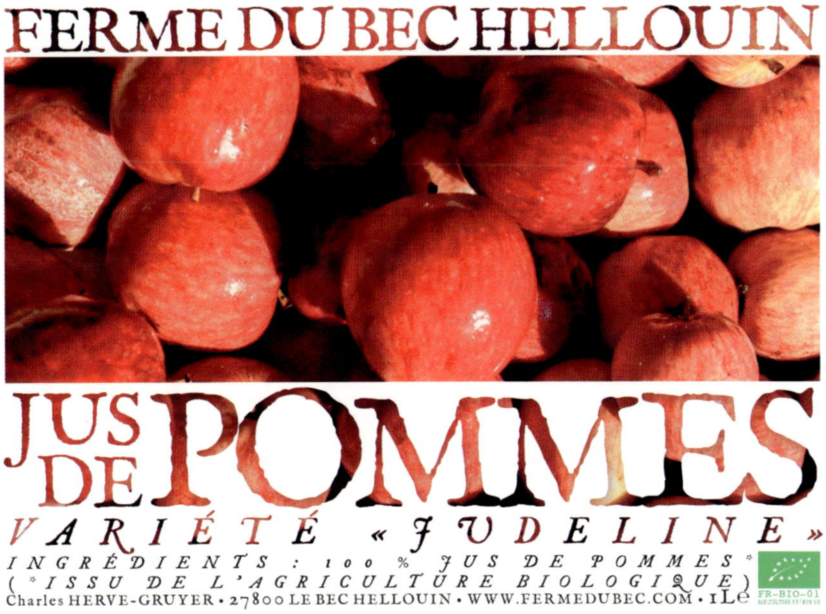

◂ *After several years we adopted a more contemporary design with the help of our graphic designer. We wanted to give plenty of space for images, but the necessary text is numerous and the labels are small, so a happy compromise was found by extending the photo into the letters. You can read some of our recipes on these labels.*

► *The obsessive quest for quality that drives our farm team also inspires our chef to make small batches of original artisanal products, created with currently available crops, week after week.*

Markets

Whether they are open-air markets, producer markets, or organic markets, these venues give farmers and consumers an opportunity to get to know one another.

Being present in a market is a real investment, and the most time-consuming way to sell. Between preparation, transportation, setting up the market stand, and making the sales, followed by return and storage, a market may require more than eight hours of work, excluding harvests. You need a year-long, regular presence to build up stable clientele. Selling at a market is practically a vocation, but those who like it wouldn't give it up for the world.

Veg Boxes, Company Baskets

Vegetable boxes/baskets have been a real craze for the past two decades. Beginning in Japan in the form of teikei, widespread in the Anglo-Saxon world under the name of community supported agriculture (CSA) before being introduced into France with the goal of supporting a producer through the commitment of a collective of consumers, who buy a farm's produce. This commitment can take different forms; it is most often an annual contract. The producer benefits from visibility and contractualised turnover. In exchange, he undertakes to share his production, with participants taking some of the risks related to weather. The boxes can sometimes be lower in value than their selling price, and sometimes superior during good times.

This formula is based on the commitment of a core group of client-partners who drive the association. Participatory projects can be organised on the farm to support the producer and allow members to speak with the farmers and discover the gardens.

CSA box systems offer many advantages to producers; it is a good formula especially in the beginning. However, this solution requires a strong investment on both sides to ensure its sustainability and it puts pressure on the shoulders of the producer who must try to honour their commitments all year round.

Contracting scares most consumers. After a few years we removed this obligation and replaced it with a lighter constraint: an order from the cart at least twenty-four hours in advance, via the internet. The number of baskets increased following this. Basket distributions can also be organised within a company. We deliver baskets to the Opera company of

Rhubarbe
Botte oignons
1,45€

Rouen, which allowed us to discover this very beautiful universe.

The Secrets of Successful Vegetable Baskets

When they choose to purchase from a small local producer rather than in a supermarket, customers search for items comprising:

- Organic production
- Local production
- Freshness
- Seasonality
- Quality
- Dialogue and proximity.

These elements can be highlighted by the producer in their communication. We don't think prices have to be off-putting; they can be aligned with those of the organic shops. The argument for cheaper prices often turns against the producer: people will not believe it's organic when sold cheap. We should be proud of our work and ask for a fair price.

Vegetable baskets should have sufficient diversity. Some market gardeners make up their basket with unsold items from the market from the day before or only offer winter squash, cabbage and leeks. This is not the way to retain your CSA customers.

The basket should have the usual vegetables used in everyday cooking and eating, and not too many uncommon and confusing products. A balanced basket is composed of two categories of products:

- The 'base of the basket': classic in-season vegetables that everyone knows and appreciates.
- The 'top of the basket': one or two original products accompanied, if possible, by a recipe. It can be a bunch of aromatics, or a heritage vegetable for example.

Organic Shops

Organic stores, although often grouped together in chains and benefiting from centralised purchasing, increasingly appreciate highlighting local producers. For the market gardener, it can be interesting to establish a partnership with a reseller. Certainly, the profits are around 25-30% lower than those of direct sales, but the time spent on marketing is significantly reduced and the volumes ordered can be large.

Restaurants

Working with one or more chefs poses challenges and constraints, but makes it possible to promote a large range of products. The working relationship operates sustainably only on the condition that the two parties are invested and communicate effectively. However, the chefs, like market gardeners, have extremely demanding jobs and are under intense pressure. The market gardener depends on the weather and seasons. The chef cannot redo his menu at the last moment to adapt... Everyone must understand the

▸ The vegetables offered in the farm shop are the same as those sold to chefs, but larger because our local customers are not looking for mini vegetables. The prices are often lower than conventional vegetables offered at nearby supermarkets.

job of the other and meet halfway. Satisfying a chef requires an ability to provide the requested products with regularity. The chef, for his part, needs to be more flexible than when ordering from a wholesaler.

There is a strong trend for restaurants to offer locally grown food and identify the origin of the products on the map.

We have worked with many of the best French chefs. Our products have also been distributed by a high-end wholesaler to the most prestigious restaurants, like the Élysée. The requirement for quality and consistency is enormous, but it is also a source of discovery. The collaboration with restaurateurs is a bit like our department of 'research and development', and all our other customers benefit from the progress made. In France, each medium-sized restaurant orders several tens of thousands of euros per year. Many more market gardeners could work with chefs in their area and build great working relationships.

◄ *Farm customers discover products that they are not familiar with, like edible flowers. It can be good to surprise them, but we always offer the classic vegetables too.*

Wholesalers

A growing number of wholesalers are looking for organic producers. Some organise a weekly collection point. Working with a wholesaler simplifies sales and logistics, but the prices are not very high.

We have a long-standing partnership with a wholesaler who orders a relatively large quantity of aromatic plants. His prices are low since he then distributes them to resellers, making two intermediaries between us and the end customer. But we can't sell as many aromatics in our other networks.

Collective Catering

Selling to a catering collective is very difficult for a small producer. The volumes needed are large, and some types of collective catering, such as school canteens, close during the holidays, especially in summer. Even when there is a desire to move towards organic and local, many chefs are reluctant to change their work habits.

Producer organisations that structure the sector, centralise supply and demand, and manage the logistics, however, allow small farms to have access to this open market growth. The objective recently set for France of 50% organic products in collective catering by 2022 represents a great opportunity.

Comparison between the Different Outlets

There is no miracle formula; each mode of distribution has advantages and disadvantages. Given the low remuneration of organic products, formulas that allow you to do without an intermediary are obviously advantageous. But the time spent on these sales can become huge!

Common sense dictates that effective marketing is based on a constant quest for quality, a sincere desire to understand and satisfy the expectations of customers, and open dialogue.

Establishing your farm as close as possible to your customers, on the outskirts of a medium-sized city for example, will have a profound influence on sales: it will be easier to sell directly, at a better price, in less time. Future farmers tempted to set up in the countryside, far away from everything, may save a few thousand euros on the price of land, but should think twice about it.

Communication

Communication is very simple, it is just about sending a message: this is who we are and what we do, the objectives that we follow. Your goal is not to attract thousands of visitors nor proclaim that you produce the best vegetables in the region, but more modestly, to attract people who share your values.

They will become customers and perhaps even friends.

The best possible communication is your products themselves: they speak for you. But realistically, one does not become a market gardening virtuoso right away. It is especially during the early stages that the farm needs visibility. Despite the initial awkwardness and errors, the farm will need this type of exposure. Your communication will benefit from being as honest as possible. In a world that is corrupted by empty commercial messages, many people will be sensitive to your integrity and your commitment to the Earth. Thanks to their loyalty over the years, the quality of your products will continue to improve and your farm will anchor itself in the local landscape.

Visual Communication

There are many different types of communication. High-quality visual communication reflects your values: packaging, labels, flyers, panels, signage, logo, website and more. The era when we tinkered with marketing ourselves online is over. We live in a world of images and are accustomed to strong visuals. Do not hesitate to call on the services of a professional graphic designer. These services are necessarily expensive. You can even barter and pay for it in vegetables. You can also solicit the talents of friends.

Event Communication

Event communication gets the word out for an event at your farm: open house, entertainment, a show, concert, a film, or a meeting-debate are a few examples. Welcoming a class or a group of elected officials is also a good way to host a successful event. Local media love to cover this kind of occasion.

Press Relations

Press relations consist of informing the media about your activities and market events that punctuate your season. To do this, send a press release to the editor describing everything about your event.

If your farm is regularly contacted by the media, have a press kit ready to avoid repeating the same thing over and over again. The kit provides essential information about your farm. It can be accompanied by royalty free photos.

The Golden Rules of Communication

Everyone knows that we live in a world that is dominated by the media. However, the goal of almost all media is to gain maximum audience to generate revenue, especially advertisement revenue. To do this, private journalists often legitimise the information that produces maximum effect in a short time. The bad news generates more emotions than good, so the media talk more about what is wrong, violence and destruction, than solutions.

The result is a disillusioned society with anxiety about the future. This is especially true for young people. But awareness is growing. Many media outlets are beginning to understand that this type of communication is an avatar of a commercial society founded

▸ *We place a lot of importance in the quality of the welcome, which goes from a smile to the floral display of the farm. Our children, who do not necessarily enjoy working in the gardens, are happy to participate in welcoming visitors. This gives them opportunities to collaborate with adults and develop self-confidence.*

on short-term profit, and that it leads us into a corner.

By committing to your project, you write a very different story: a story of courage and risk in favour of the common good. You look for solutions to avoid the chaos and invent a better world. Media outlets are looking for this kind of story.

Here are the strong elements that will make your farm outstanding:

- The place is beautiful and exudes harmony: the journalists are in search of beautiful images.
- Your personal trajectory is profound: the journalists wish to support their story with an authentic human adventure.
- You are looking for meaning and solutions: all over the world, people are on a similar quest; what you discover will interest them.
- You agree to make yourself available, you are open to sharing, and you know how to make people feel welcome.

When the media relays this type of meaningful story, it makes the general public aware of solutions, gives hope and arouses a desire to get involved. As you know, the biosphere is degrading so quickly that we have very little time to reverse the downward curve. What can each of us accomplish without means or relationships? Well, one answer is to use the media and their incredible ability to touch so many millions of people. Communicating is a way to strengthen the economic viability of your farm, on the one hand, while participating in the transformation of the contemporary world, on the other. In the field of marketing as in others, what really matters is that you are comfortable in a system that is connected through communication.

▲ *We welcome the media because the stories they share show the values that we support. These types of stories can become powerful accelerators of societal transition.*

◄ *Shanti with actor and director Mélanie Laurent, during filming of the documentary* Demain, *that Mélanie co-directed with our friend Cyril Dion. This award-winning film was shown in many countries. It describes a new world, made possible by those who become involved locally.*

Parking

◄ We first opened the gates of our family farm in 2007 and were surprised by the expectations of visitors. For many of them, visiting a small farm was like drinking from a mountain spring. We could see that the societal aspects of a farm outweigh its economic impact. It produces much more than food!

20

THE ECONOMICS OF A MICROFARM

I grew vegetables and produced food for people and animals for forty years, and the superiority of vegetables grown according to good organic practices is obvious. There is as much difference between one of our carrots and the ordinary supermarket carrot as between a Rolls Royce and a Citroen deux chevaux.

Eliot Coleman[85]

You are now convinced: a microfarm is a business. Good management is as necessary in the agricultural sector as in other sectors.

A Good Expert Accountant

When you create your farm, you will be looking for the best varieties of fruit trees and the most suitable breed of chickens. Employ the same level of scrutiny when choosing an accounting firm. It is difficult to master all of the skills you need on a farm, and the services of a good accountant are a necessity. The accountant is a partner who must be able to reliably answer questions and help you to choose the best options.

We learned that a small farm with little income is not a desirable customer for certain firms. It resulted in subpar service and bad advice so we changed firms. Do not hesitate to be demanding; your farm deserves it.[86]

Invest a Little or a Lot?

In your initial investment choices, be rational. Better to start slowly and gradually build up. Money does not come easily in the first few years; undoubtedly it is preferable to keep some savings in reserve rather than invest everything at once.

Avoid borrowing money. Repaying a bank when you don't get enough margin to pay yourself can be deeply discouraging.

For his doctoral thesis, which extended our study, *Permacultural Organic Market Gardening and Economic Performance,* Kevin Morel studied a group of small farms and identified three categories of strategic choices:

1. The **technical system** has three distinct approaches:

- o **Manual micro-agriculture (Mi):** 100% manual, high densification, crop associations, low-input practices (compost, mulching, natural preparation and more), no green manure, 4-6 crop successions per year.
- o **Biointensive market gardening (Bi):** medium mechanisation, high densification, no crop associations, low-input practices, green fertilisers, 1-4 crop successions per year.
- o **Classic organic diversified market gardening (Cl):** high mechanisation, low densification, no associations, no low-input practices, no green manure, one or two crop successions per year.

2. **Sales**

- o Over 12 months of the year with production of low value storage vegetables.
- o Over nine months of the year without production of storage vegetables.

3. **Investment:**

- o Installation without equity at the start:
  - Low-cost installation: second-hand equipment, recovery, self-construction.
  - High cost installation: new equipment, not self-construction.

▼ We should not underestimate the time spent on sales. Being present in the markets is very time-consuming and it takes months, sometimes a year, before building customer loyalty. Markets offer visibility and promote a direct relationship between producers and consumers.

- Routine, five years after installation.

Kevin combined these different possibilities to make 18 scenarios that he built into his computer model to identify the conditions of economic viability (maximum income and minimum working time). These results emerged:

- Intensification per unit area and low-input ecological practices (Mi and Bi) enable a higher level of sustainable economics compared to the traditional approach. The slightly mechanised approach (Bi) is more efficient than the fully manual approach (Mi) (let us point out, however, that the market gardeners studied by Kevin do not use the same method as our farm described in this manual, which allows you to reach higher economic efficiency).
- Intensification per unit area and low input practices (Mi and Bi) resulted in a superior level of economic viability compared to the classic approach.
- Marketing over nine months without conservation crops increases the chances of economic viability compared to a 12-month marketing strategy with production of conservation vegetables.
- A recently installed system is less viable and more fragile than an established system that has reached the routine phase.
- At first, a less motorised system on a small area is more viable than mechanised: fewer loans for equipment and land.
- Low-cost installation reduces the chances of viability: the work overload linked to self-construction of tools does not compensate for the reduction in borrowings carried out, in a hypothetical installation without equity (again, let us specify that the farms studied do not use the Bec Hellouin range of tools, including the campagnole, which was not yet commercialised during this study).

Ultimately, Kevin identifies several levers that increase the chances of economic viability upon installation:

- Have equity.
- Use aid: start-up funding but also regional aid...
- Take advantage of a collective solidarity, sharing and bartering.
- Have other sources of income.
- Have several marketing channels (Kevin's study only focuses on systems that have weekly CSA baskets).
- Benefit from technical innovations linked to citizenship of farms.[87]

▲ *Our dream would be to sell all our production at the farm but our remote location in the countryside does not allow it. In this photo the discreet, wonderfully effective Edith, who has managed the farm for six years, is hiding behind Fénoua.*

► *Shanti and Fanny welcome visitors and give them an explanatory leaflet. Entrance fees make it possible to finance part of the maintenance costs of the farm.*

Creating a Microfarm: How much Does it Cost?

Estimating the budget necessary for getting started is both essential and difficult: personal choices and conditions of installation vary from person to person.

During our last training, a young man gave us a tour of his farm that he established four years prior with his companion. They began in a minimalist way, having only 20,000 euros patiently saved during the years preceding their installation. No borrowing, no subsidy. They were able to buy once hectare in an isolated countryside setting – woods, hedges, stream – with a ruin that they restored over time.

They live in a yurt without running water or electricity. They have no mechanisation and produce vegetables in permanent beds that are prepared with a campagnole. All their produce is sold at the local market. Believe it or not, this couple has been living off their farm since year one, reinvesting 10,000 euros per year, and they even hired an employee.

We also met market gardeners who had changed their occupation after a successful career and immediately invested 200,000 euros to buy a farm (excluding land).

This is a very rough estimate of what a microfarm in France can cost currently, not including a dwelling.

An initial investment budget of this order allows you to start with a microfarm that is well structured. The equipment and tools category is cheap compared to what it would cost for a motorised farm. Money saved on machines is invested in fruit trees, hedges, and ponds.

If you do not have this amount of money and you want to avoid resorting to loans, it is entirely possible to start with just the land and a minimal amount of equipment, with no greenhouse, no planting of fruit trees or hedges, without ponds and without paths.

It will probably be more satisfactory to start modestly, without generating too many constraints with planting work, construction, and infrastructure. You can rent a mini excavator and do the earthwork yourself – digging a pond, planting holes; it will save a lot compared to hiring a contractor.

A realistic budget, from our experience, falls between 50,000 and 100,000 euros.

Purchase of agricultural land of 1 hectare (notary fees included)	€10,000
Fences and gates	€5,000
Equipment and tools, equipment for marketing	€22,000
Tool shed	€10,000
Used vehicle	€5,000
Planting an orchard and a hedge	€10,000
Digging a pond	€5,000
Paths inside the land	€5,000
Office equipment	€2,000
Subtotal	€74,000
Miscellaneous (10% of subtotal)	€7,400
TOTAL (all taxes included)	€81,400

◄ *Although we work outdoors, with our dog on our lap, managing the farm is no less serious than other types of businesses.*

◄ *The construction of the ecocentre was a crazy adventure, completely disproportionate in time and budget. This bioclimatic building, built with reclaimed and natural materials, was constructed entirely by local artisans. It took three years to complete. Finally, we can welcome our trainees in a beautiful setting.*

21

BUILDINGS

The architectural organisation of farm buildings differed in the past more than today, due to materials that were specific to each geographical region. The builders of the past did not primarily seek to create architecturally pleasing buildings. They adapted their structures to the various activities of the farmer, his family, and animals. They had knowledge and practices that were developed, carefully considered and experimented over generations that allowed them to practise their craft.

Marie-Claire Ricard and Hervé Monestier[88]

When you create a farm, your time and attention is monopolised by crops, tree planting, and acquiring equipment. For many, the question of buildings comes later. They make do with fewer resources for construction and end up with poor quality, unaesthetic, high energy use buildings that lack functionality, and don't collect rainwater.

Creating a farm should be cross generational. It will have a strong impact on the surrounding area, your quality of life and the experience of your visitors. I am extremely passionate about rural buildings, and my vision is necessarily subjective. We invested a lot of time and money in buildings. A new one every year is obsessive and unreasonable, I agree. I exhausted Perrine and the team, and spent all the money that came in, but this quest for beauty and anchoring in the land allows us to recreate the functionality and charm of the farms that I knew in my childhood. The difference is that we realised in just 15 years what the peasants of the past learned over several generations: the grandfather built the cottage, the father the barn and the stable, the son the henhouse and the walls around the garden... Millions of these farms have been destroyed over the last 50 years, or transformed into second homes. What a tragedy when we are forced to return to working the land: will we manage to rebuild this lost legacy?

Beauty is unreasonable; our investment in building our farm certainly is as well. But is life worth living without beauty?

► *Children are introduced to laying cob, a mixture of clay and straw, between the half-timber, wooden joists that form the structure.*

Farming Architecture around the World

The small farmers of yesteryear, just like the fishermen, were poor. They made do with what they had: wood, stones, pebbles, earth, sand, bamboo, clay, and lime that they could collect from their land. These peasants and sailors, helped by village artisans, transformed and

▲ *The structure of the first part of the ecocentre is finally in place! The south facade is almost entirely glazed. This building will retain heat very well when it is finished.*

assembled natural materials with their hands and some simple tools: axe, adze, hammer and some others that they made themselves. This resulted in modestly sized buildings and boats.

As they had no source of energy beyond their strength and that of their animals, or the force of the flowing river and the blowing wind, to operate their equipment, each tool was adapted precisely for its use. Weak energy is too precious to waste.

The use of natural materials and manual labour resulted in buildings, sailboats, and tools that were simple and beautiful. They were perfectly connected to the landscape and seemed to flow from the earth. They were also inexpensive and very effective. We could add that they were in line with the limits

of the planet. This is true in all cultures of humanity. Whatever the materials and shapes, they seem right in their environment. Look at the marvels that were the peasant houses of Japan, the isbas of Russia, the log cabins of the American West, the mud houses of Africa and China, the pae pae of the Marquesas Islands...

Valuing Local Organic Resources

Most of the buildings on our farm were built with trees and stones found here, and local clay. We deliberately sought an anchor in the rural tradition of Normandy, respecting the style and proportions of the region. I continued, more or less consciously, a dream to revive a little family farm on a human scale, a happy farm as in children's books. I felt an obligation to do this correctly. However, we were not locked in the past, and also used contemporary materials like bay windows for natural light and excellent insulation, renewable energy and more. The ecocentre on the farm is a bioclimatic building.

A great archaeologist and eminent historian took several training courses at Bec Hellouin Farm. He told us: "When I came for the first time, I thought I was entering a Merovingian farm." Perhaps it makes sense that a post-oil agriculture farm looks like a pre-industrial age farm? We constantly seek to take the best from the past and the best of the present to invent the future.

This is all incredibly entertaining. Each project excites the team. We like to cut down the trees, bring them back to the farm, shape

▲ *In 2018 we built a Native American sweat lodge, under the direction of our Colombian friend, Vanessa Correa.*

◀ *Vanessa is laying the cob on a light structure of hazel and wicker. The exterior is formed with a clay coating of lime and cement.*

▲ *Squaring timber that was harvested from our forest with a doloire, an axe from the Middle Ages, is very satisfying. It is the best anti-stress activity and great for losing weight. The barn was the first building frame produced in this way.*

them, assemble the framework, and erect the structure. Jean-Claude, with the team for eleven years, participated in all of these projects. He often tells me, "I keep coming because I never get bored; I always learn something."

The Barn

One project leads to another. When we started making petroleum-free cereals, a barn with a threshing floor became necessary. We built it ourselves. The framework was cut with a doloire axe. Learning to square a trunk with this ancient tool required some research. We also had to learn to sharpen the blade perfectly. Working like this allowed us to develop some new skills.

This modest barn gives us a lot of satisfaction and we are proud to have built it together. Constructing such a building also builds team coherence.

Visitor Impressions

I don't know if this search for simple beauty, natural materials and work style amounts to a neurosis or a quest for the absolute – probably a little of both – but still, after years of effort and passion, the farm began to resemble our inner vision. We were very surprised to see the impact on our visitors. Almost everyone expresses a feeling of wonder and well-being: "It is even more beautiful than in the movies," say smiling, starry-eyed visitors. We understand that our dreams are similar to the dreams of many people. Our collective unconscious is likely inhabited by farms on a human scale that intimately connect humans, animals and plants. What we accomplished for ourselves impacted the visibility and success of the farm. In our first year, the farm was awarded an ecotourism award for Upper Normandy, a true testament to the degree that our dream resonates with many people.

Functionality

There is no opposition between beauty and functionality, quite the contrary. Constructing our own buildings with natural materials and simple means requires a lot of effort, so we

► *To cut a beam straight, start by tracing equal size squares at each end. The corners of the two squares are then connected by a line.*

► *Notches are cut up to the line, then we quickly remove large chunks of wood before making a more precise cut. The beams can also take the natural shape of the tree.*

Dayton axe

Half bevel–framing chisel

cannot afford to build useless square metres. Maximum efficiency must be achieved: 'one element, several functions'. Most of our buildings serve multiple uses. The shop, for example, where the team takes their meals, is also a kitchen where our prepared products are created, and a cider cellar. A large office space is used by our trainers for overnight stays. There is even a sauna.

Japanese chisels

Cost

If cost is the determining criterion, you might as well opt straight away for a corrugated iron building. It may be a farm that looks like a small factory, but it won't cost much. Humans are the only animal species that massively uglify the surface of the Earth.

If you want to combine low cost, functionality and beauty, use local organic resources and build things yourself, you will have to learn the right skills and techniques and will probably want to delegate certain tasks like electricity and plumbing. But there is value in doing things on your own, even from an economic point of view. If you have a budget to build a building for your farm, is it better to hire a construction company that will charge you at least 40 euros per labour hour excluding taxes while you do market gardening for only a few euros per hour? It is entirely possible to keep your money and build your own structures if the adventure tempts you, even if it means starting market gardening a few months later. You will be proud of a building that you built with your hands and you will

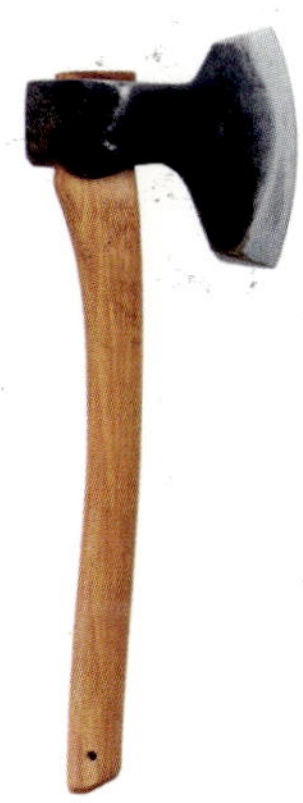

Carpenter's axe

Mortise axe

◂ *The assembly of the frame is a great day, so much emotion! The building seems a lot larger than when the frame was laid out on the ground (above).*

▸ *Perched on the bank of the Bec river, the barn was built entirely by our team. This achievement has developed a sense of pride in everyone, that we are capable of making such a pretty building by hand, from the cutting of trees in our forest to making shingles for the roof. The image that opens part XV (p.108) shows the interior of the barn, which became my workshop.*

have half the money left in the bank. You can also look for an intermediary approach and hire local craftsmen to assist you.

Pioneer Spirit

There is something of the pioneer spirit in this approach. Creating a farm yourself is an adventure; you have to be able to take risks and move beyond your old habits, but how invigorating! Daring to do big tasks builds self-confidence. Too often we doubt ourselves and become our own worst enemies. The years spent at school with rote and passive learning have eroded our ability to tackle the unknown. We do not seem to realise how much talent and skill we possess. We could achieve almost anything if we dare to take the plunge and love what we are doing.

Planes

Joiner's planes

Japanese planes

The Farmer's House

Should we live on our farm or not? Our experience drives us to say yes, for the constant monitoring of animals, greenhouses and the site, not to mention the pleasure of being there! But we must be prepared if the farm begins to attract many visitors. Too many guests could feel intrusive. If you live onsite, make sure to properly separate your personal crops from the rest of the farm. It should have a private entrance. If it is impossible for you to live on your land, give up any ideas of animals that require a daily presence – this is not the case for all. A farmyard for example can live a few days without supervision with automatic feeding and water. Include a self-closing door to protect it from foxes.

This chapter reflects our personal tastes but it is incomplete – we could fill an entire book just on farm buildings. You can use permaculture design protocols for ideas.

In a world dominated by the short term and a constant quest for immediate profit, it takes courage to spend the time and means to make beautiful and lasting achievements. This may mean planting oak and other large deciduous trees. Farms don't function with the frenetic rhythm of our time. They are built to provide shelter for many future generations. This gives meaning to our efforts.

◄ We built this bread oven in 2007. Its construction was pretty straightforward. A world opens up when doing new things like this, but a leap into the unknown is generally rewarded. This is how we learn, from project to project.

ATTENTION
ABEILLES

◄ *In the past, the agricultural economy was the backbone of the community, made up of numerous actors whose activities interacted with one another. The spirit of permaculture invites us to imagine new types of interactions between farmers and other local actors, for the benefit of all.*

22

MICROFARMS, MICROFARM ECOSYSTEMS, COMMUNITY-SUPPORTED AGRICULTURE

Is there a non-violent economy that involves no pressure and lends itself to no abuse?
A non-violent education of children and a teaching of non-violence to young and old?
A non-violent authority that does not rely on force and does not involve any privilege?
Non-violent justice, justice free from punishment or punishment free from violence?
Non-violent agriculture and breeding?

Lanza del Vasto[89]

In the globalised processes of industrial agriculture the agricultural operator is only a link of a long chain whose work contributes to the destruction of Earth, the only known living planet.

On the other hand, ecocultivators seek to do everything that can be done to create islands of fertility and biodiversity. The impact of a single small farm is local, but if tens of millions of microfarms operated on all continents, using the practices of the ancient world, then it becomes possible to heal the Earth, maybe even faster than we imagine.

When ecoculture replaces industrial agriculture, our societies will be profoundly transformed. Many assets will return to the earth. In our previous book, we mentioned the fact that 5.5 million agricultural jobs have been destroyed in France since 1950, the same as the current number of unemployed. Ecoculture could help with problems of employment, health, and precariousness - so many problems that our 'advanced' societies seem powerless to solve.

The current economic model is based on globalised and virtual finance that is totally disconnected from the real world. A circular economy, imitating the functioning of nature and based on the production of local goods - food, energy, fibres, biomaterials - is essential to human survival. This approach is better geared towards stability, security and fair distribution of resources that human beings need to flourish.

True Sustainability

The only solution that allows humanity to sustainably protect the Earth is to base production of essential goods on the value of renewable energies and biological resources, through local laws. Let us remember that:

- **Energy from the sun is clean and easily extractable on a human timescale.** The sun causes wind, hydropower and biomass, and ultimately animal and human energies;
- **The only resources capable of being renewed to live and grow in a finite world are the living resources.**

Let us keep in mind that all sophisticated technologies make use of non-renewable resources and are a source of pollution (this is the case with solar water heaters, wind turbines, photovoltaic panels, etc.). In reality, none of our so-called technologies are actually clean. The best tech alternative is an alternative to technologies!

It is better to harness the sun's energy through bio-processes rather than use technology, and plant trees that naturally store carbon through photosynthesis, and use this wood to produce energy rather than use a wind turbine. Wood can also allow you to build a house, make a fence, tools, dishes, and toys. Tree leaves create humus, feed livestock and can be used for healing, all things that a wind turbine cannot do. Making use of all that

▲ *Recreating connections between breeding activities and crops improves the resilience of farms and local food security.*

trees can offer can only be done by the people who live locally, like farmers and artisans.

In our vision, industrial structures will eventually be replaced by workshops that are designed at human scale. There will also be an increase in the number of artisans. **There is a deep and natural connection between the workshops of small farms and artisans.**

Low-tech solutions are the most durable. This does not mean that we should do without all that high tech has to offer. The internet is, for example, a tool that is helpful with work aimed at protecting the biosphere. We should use technology with discernment to develop solutions that have no negative impact on the environment.

Farmers and Artisans

Microfarms make the emergence of human scale communities possible. These communities value the biological resources of their territory across networks of exchange, and the solidarity that closely links farmers and artisans. Their activities support retailers, service providers, and educational, cultural, therapeutic and spiritual institutions.

These emerging communities will resemble those of the pre-industrial era, however their operation will be based on the incredible amount of knowledge acquired during modern times about the functioning of nature and the human population. They will be connected to all other local communities of the world through evolving systems of communication. This will not be a return to the past.

Although our planet is finite, we can invent ways of life that are sustainable and allow everyone to live with dignity. By eliminating waste and through efficient use of goods and materials, it will be possible to eradicate poverty. Intangible things like art, culture and spirituality are unlimited and can be widely shared among all. Cultures will thus be enriched along with their respective treasures.

The current crisis compels us to embrace a holistic vision of the planet. Threats to our survival will gradually awaken an awareness that the supreme good is nature and life itself. This is an opportunity to invent new ways to inhabit the Earth. Our relationship to nature is at the core of this transformation. Small farmers will play a decisive role in the passage from one world to another, acting as the interface between the human population and nature.

This vision may seem utopian, but humanity today consumes more than a planet and a half worth of resources, while we are 8.1 billion human beings (as of June 2024). In 2100 the planetary population will be around 11 billion. How many resources will it require then? What is actually utopian is the belief that there is a planet B on the horizon. We urgently need to find solutions to live in harmony with our unique and precious Earth.

Changes to Come

One of the most exciting aspects of a microfarm is that there is no need to be rich and highly educated to be a market gardener. Anyone can do it starting today. With hope that these individual actions will have a strong environmental and societal impact.

In this manual, we have focused our attention on microfarms. However, we do not imagine that it is enough to depend on a single farm as a model – it would be sad to have just a single model or one approach to farming.

▶ *Being isolated, we had to invest in all of the necessary equipment. It would have been so much less burdensome to share a draught horse and tools, a shop or a common secretary with other farms.*

In many parts of the world, farms are large, some even very large. We will have to imagine their transition. The concepts of permaculture, even if they are better suited to small-scale structures, can provide valuable insights.

In our previous book, *Miraculous Abundance,* we described two versions of the microfarm concept. We will discuss them briefly here and invite readers interested in this theme to refer to the book.[90]

▲ *Farms create social links. It is easier to host when we can distribute the production and reception tasks between several people.*

Microfarm Ecosystems

Microfarm ecosystems consist of a group of farmers and artisans with complementary activities, a market gardener, an arborist, a breeder, a beekeeper, a producer of aromatic plants, for example. A 5-10 hectare plot that would not accommodate a single farm with current practices, could host all of these activities.

The diversity of a single microfarm would potentially be a lot richer within such an ecosystem. The connections between these activities would be many: the breeder's manure would fertilise the land and the animals would benefit from crops. The beekeeper's bees would pollinate the crops, for example.

This association would make it possible to pool a certain number of expenses like a website, administration, equipment, and a shared point of sales. The microfarm ecosystem store would be well stocked with various products and open more often than a single farm shop, which would allow each producer to benefit from better marketing at lower costs.

Such associations could have separate legal status - we are in favour of an association of owners each managing their own activity, to limit the risks of conflicts inherent in a community.

Solidarity Agrarian Systems

The solidarity agrarian system is based on the vision described throughout these pages, but concerns a larger territory. The idea is to produce everything the region is capable of producing, including grains, meat, dairy products and more, in order to diversify these activities and promote food security.

The Farmers of Tomorrow

The spirit of permaculture is an invitation to adapt each establishment to its territory and to the people who live there. There is no single answer: there are as many answers as there are project leaders. The creativity deployed and the diversity of solutions explored will lead to collective progress.

It seems highly probable that the farmers of tomorrow will not come from farming families since this profession has become so compromised and diminished. The people who work the land in the future will be driven by a desire for reconnection and commitment. They will be in search of meaning and will not practise agriculture in a manner that destroys soils and biodiversity.

Many teachers in French agricultural schools do not understand the aspirations of their students, resulting in tension between faculty and students. A former student sent an email recently and described a problem that has been repeatedly spoken about: within the entire class of fifteen students, fourteen aspire to create a permaculture microfarm. The teachers don't seem to believe in this model... The world of organic agriculture has not yet realised the problems created by its dependence on fossil fuels, nor the destructive impact of mechanisation on soil. We can see that the problem is changing rapidly in the right direction. It is helpful to keep this momentum by displaying kindness, without rigidity or discouragement, without being overcome by obstacles. We are not in a battle with anyone. Let's not waste our energy over controversies; it is better to devote ourselves entirely to developing alternatives that are so effective and appealing that they will win people over more effectively than all the oratorical jousts of the world.

▼ *Working with a team allows you to share tasks.*

◄ *When we were asked to offer training on the farm, we could not have imagined the adventure we were starting. Facilitating training is a total commitment which can feel overwhelming, but what an amazing group of people we have met. The first few years taught us a lot, especially due to the people we've met. Training weeks now feel like time spent with friends.*

23

OFFERING TRAINING AT YOUR FARM

It is time for humanity to realise how closely they are linked to the Earth and the cosmos.

They take, they take, they never stop taking, forgetting that the Earth is a heritage that must be managed for the good of all, for those who live today but also for those who will come tomorrow and for countless generations to come. The Earth is like a garden that must be taken care of thoughtfully, being careful not to exhaust the soil. All the harm we do to it comes back with a boomerang effect.

Dhyani Ywahoo, Etowa Cherokee[91]

▼ *Perrine and I take great pleasure in sharing our passions. The confidence of those who come to train with us puts pressure on the whole team to constantly improve the level of our craft. It is like walking a fine line, with each misstep possibly causing a trainee to drop out.*

The transition from agriculture to eco-culture will require professional training. According to Pablo Servigne,[92] in a report commissioned by the European Parliament, no less than 117 million new farmers will be needed to cultivate Europe by 2030, to ensure food security on the continent in the event of an energy crisis.[93] To calculate this astounding estimate, Pablo Servigne relied on his experience from Cuba. The island was highly industrialised when the fall of the USSR caused a brutal collapse of its economy. In less than 10 years, 25% of the Cuban population began to work the land to feed the island.

Schools of agriculture in France are not prepared to fulfil the need for training in natural agriculture. They lack familiarity with the subject and practical experience in the field. The institutions in France are aware of this and are beginning to qualify teachers for these emerging themes. We have been asked to contribute to this effort. The speed with which agricultural education is adapting is astonishing and deserves praise, but there is much more progress needed to satisfy this urgent demand.

There is an immense need for human resources and land to cultivate. Our farms could play a major role in training the farmers of tomorrow. Practitioners who possess the ability to teach are the best candidates to convey ecoculture techniques. However, be careful not to overdo it, by creating complete market gardening training that could overwhelm farmers. Instead, it would be wise to try out various forms of partnerships between agricultural education establishments and farms. Farmers can offer specific training and practical internships, but also partake in course development at the agricultural schools. The schools take responsibility for the programmes, all general subjects, logistics and financing.

▸ *The large room in the ecocentre is a comfortable place for plenary sessions. The large number of trainees can be daunting at the beginning and impose some constraints, but we quickly notice that the meeting of so many enthusiasts who come from diverse backgrounds and countries is very rich.*

Become a Trainer

Training future professionals requires specific skills. You need a profound depth of knowledge in the subjects taught, to have structured thinking, know how to explain things clearly and vividly and enjoy sharing the knowledge. Pedagogy is a profession and an art if we aim for excellence.

Farmers who become trainers have the legitimacy factor of living what they teach, but they must cultivate the skills that make a good teacher. One can be the best of farmers and a poor trainer. Conversely, excellent teachers do not necessarily need daily practice in their subject matter.

We could add that the practitioner who aspires to teach will be the first to benefit, as students motivate you to broaden your knowledge and clarify your ideas.

▾ *Hands-on workshops are made up of small groups. The wide diversity of activities practised on the farm provides lots of different apprenticeships, all located within a very small area.*

Offer Training on your Farm

There are many advantages to organising training sessions at your farm. Your motivation to master the skills of market gardening will skyrocket, you will experience gatherings of wonderful people and benefit from a source of additional income. The public will have the opportunity to gain experience at an innovative farm and train with a real practitioner.

Preparation for training can take several years, and involve many components:

- The site must illustrate the teaching offered and serve as support for practical workshops. This may require some adjustments to your farm.
- Suitable premis is essential to accommodate the public.
- Trainers must be professionally skilled in the craft and meet the educational requirements.
- It will be necessary to create additional teaching materials.
- Offering training classes entails legal formalities, promotion, administration and more.
- Training logistics include reception, meals, accommodations and more.

Grow in Power

The preceding points demonstrate that offering training at your farm should not be done haphazardly. While some folks may be self-proclaimed trainers, keep in mind that mediocre training has little value.

The amount of the work necessary to be able to offer quality training should not be an obstacle. If we have the desire to share, there is no reason that we can't become effective trainers. Common sense suggests starting out small. Organising introductory gardening weekends for around 10 trainees is relatively simple. Your farm's infrastructure should be able to accommodate a small group, both for classroom sessions and for practical workshops. A simple barn can be very suitable for classroom sessions and mealtime. You won't need a full-time secretary and you won't necessarily need to create additional teaching materials for such a short course. Finally, any market gardener who has a few years of practice is well equipped to transmit the basic knowledge needed for beginning gardeners.

If we honestly and consistently analyse what worked well after each training session and make improvements where needed, we will see continuous progress. Starting small allows you to test yourself and decide if you are motivated enough to ramp up. If so, the experience acquired will allow you to offer increasingly advanced training. You can supplement your own knowledge by inviting expert speakers.

Feedback on Training Sessions at Bec Hellouin Farm

Here are some thoughts about our experience. We were asked to organise training for the general public in 2008, two years after our opening. We didn't have any problems doing it because we love to teach and share our passion. Moreover, teaching about the environment was my previous profession.

We began offering a professional market gardening course in 2010. It is a 35-days course with 35 participants. The first big training course combined the farm team with professional teachers. Several celebrated pioneers of organic farming agreed to get involved. Some of these initial students joined our permanent team. Training at Bec Hellouin Farm was off to a good start.

Since then, training has been managed by the Bec Hellouin Permaculture School, a legal entity separate from the farm. It never stopped developing and quickly required a full-time secretary, and a chef. Its delicious organic and vegetarian meals, cooked with farm products, go a long way to highlighting the success of the school. We also have built the ecocentre to host this activity in the best conditions.

◂ *If training is 100% full year after year, it is undeniably linked to the excellent organic meals and vegetarian dishes that chef Fabien concocts.*

Unable to satisfy all requests, we stopped our general public training in 2016 and only offer professional training aimed at farmers and future farmers, in three cycles of five days: very intense, focused on micro-agriculture as we practise it. These internships come in addition to official agricultural training. We have also introduced internships on interior ecology, a topic that is particularly close to our hearts. These training courses allow us to pacify our inner world, thus creating a more favourable context for a peaceful relationship with the outside world.

For several years now, all our training sessions are full and the waiting list has hundreds of people on it! This means the demand is significant. This poses a problem of chronic overload for our secretary.

▾ *The meeting of so many explorers of the future in the same room is lively. The courses are so intense that strong bonds are formed between participants that often endure, in some cases leading to marriage and children.*

▸ *A training session is at once an immersion in nature and humanity, a mix of ideas and emotions of rare intensity. When Rose decided to train in permaculture, at sixteen years old, she saw the family farm in a new light. Today she is a student of Fine Arts, and intends to design edible landscapes.*

The numbers increased gradually. We limit it to 50 trainees, despite the demand. This may seem enormous, but the progressive increase in staff numbers was accompanied by the creation of a very reliable team that can accommodate a large group. Each training links trainees to a network of enthusiasts who share skills and help each other.

At the end of each course, participants are asked to complete a satisfaction questionnaire. The questions cover all aspects and all sessions. These questionnaires are examined with great care by the team. In the early years, some issues became apparent. We took this into account to constantly improve training. This has helped raise the level of satisfaction, which reached 100% for several years, except for rare exceptions. But we continue to pay close attention to these surveys.

Perrine and I learned a lot along the way. Having to teach regularly pushes us to constantly work to improve our course. I'm trying to correct my tendency to dilute the contents too much and go off in different directions. I noticed that at certain times I manage to arouse emotion, laughter, intense concentration, and I seek to develop anything that helps trainees focus their attention and learn while having fun. Teaching is also a bit like acting. It helps me heal my shyness. I must admit that I now love being face to face with 50 strangers at the beginning of a training week. Each session is a journey during which we must unite all these unique individuals and lead them, for five days, through a vast field of knowledge. All while having a good time together.

We always remind the group on opening day that the succinct training depends largely on what each one gives of themselves. Training of this type is not a product that you buy and that we consume, but a human adventure that we construct together.

We seek to offer training that we would have liked to receive ourselves; training that saves years of hard work and is economically rewarding. We have chosen to increase the cost gradually, preferring to seek out the best possible quality in everything, which has a cost. However, training can be supported by different organisations and social diversity is real. The tuition also finances part of the research programmes that take place on the farm, which have no state funding. Our interns are the first beneficiaries of the data collected during these programmes.

Ingredients for Successful Training

Training is like a cooking recipe: you need a certain number of plentiful ingredients so that the final dish is tasty, neither too bland, nor too spicy. Here are some of the ingredients that enable training to be successful.

Work hard, neglect nothing: programme quality, timely answers to questions, efficient registration process, a beautiful environment, delicious high-quality meals, plenty of tools that are well organised in the workshops, clean toilets and much more.

Welcoming everyone: for many trainees, staying overnight on a farm comes with a bit of anxiety. So, there is always someone in the parking lot to help them park and find their way, a person who accompanies them to the ecocentre and helps them settle in, and offers a welcome coffee. Throughout the week the facilitators make themselves available so that no one feels isolated.

Listen to them: trainees may not understand this or that point, or they can be disappointed or irritated by an aspect of the training or by a certain behaviour. We invite them to express their questions, criticisms and suggestions without holding back in order to keep problems from getting worse. There are different ways to facilitate this dialogue, including a system of 'tiny notes' for those who would like to remain anonymous.

Be responsive: each morning our teachers review the previous day's teaching, and prepare for the one to follow. Small adjustments are made day by day; course comments help us progress thanks to the view of others.

Programme: at the start of the course, each participant is given a programme explaining the objectives, and the methods to be used. At the start of each day, we review its contents, an agenda that can change due to bad weather.

Classroom: the demeanour of trainers is important. Students must feel like the teacher knows their subject thoroughly, that they are sincerely invested in teaching and that they are happy to be there.

Sessions must be solidly grounded with a coherent plan. However, they must not fall into a rigid rhythm. Anecdotes, humour, stretching and breathing sessions keep minds fresh.

The trainer regularly ensures that everyone is following along and understands. They must keep the thread of the session focused and leave space for questions and feedback from participants. In case of a runaway talker, you have to know how to tactfully redirect anyone who would otherwise monopolise the floor.

Course materials: classroom lessons are often accompanied by a slideshow. It should not have too much text. There is nothing more boring than a trainer who reads the text of his slides. The beauty of photos, diagrams, tables and illustrations help to capture attention. Each trainee is given an additional booklet of around a hundred pages that summarises each theoretical and practical session, with spaces for taking notes. It also offers bibliographic sources for further study. This saves participants from having to write everything down and frees up their attention.

Practical workshops: field experience is essential. In training intended for professionals, fieldwork is less than with general public training and we try to focus on themes that they have little or no opportunity to practise on their own farm. During practical workshops, the groups are made up of one or two facilitators for around eight participants. The sessions generally last two hours.

Trainees must not feel like they are producing actual work for the farm, like they actually paid to work for us. However, just listening to us talk without getting involved is not satisfactory – they miss out on their opportunity for hands-on experience. There is a balance to be sought between these two extremes.

▲ *Teddy teaches students how to use the tilther, a tool designed by Eliot Coleman.*

▼ *What happens among trainees outside class and workshop hours is important and contributes to the richness of the training. The river and pond are a great setting for hosting festive times.*

▲ *The diversity of encounters and the mixing of ideas often give rise to important questions and perhaps a few nights of insomnia. Training is a time apart, a break that we take, a distancing from daily life that can lead to new life choices.*

Balance between theory and practice: one key to successful training is to be able to learn with your whole being – your body, your mind and your different forms of intelligence and emotion. Every day involves several hours outside, usually the first half of the afternoon. When the weather is good, many activities take place outdoors. This allows one to breathe and be immersed in nature.

A balance between lectures and open exchanges: there is so much to share in five days that we offer very intense training, from 9am to 7pm, and often a session after dinner. To cover the programme, we are required to offer a certain number of plenary sessions during which teaching is done in a traditional classroom style. This could induce a certain passivity of the trainees, but we leave lots of space for questions and feedback.

Participants can present their achievements in the evening. The plenary sessions are supplemented by small workshop groups, sometimes in the form of a game, with a presentation by each group at the end of the week. What happens between the trainees outside of sessions is as important as the lessons.

For years, we have livened up the morning with a review, a show of about 15 minutes summarising the previous day, in a humorous manner by a group of trainees. This bonds the group while giving rise to countless hilarious sketches.

▶ *The fact that the training takes place at a real farm, and not just a simple demonstration garden, is relatively exceptional and contributes significantly to the quality of learning. Let's hope that, in the near future, our land will be criss-crossed with farms open to those who are eager to learn.*

Sharing experiences: training at our farm benefits from solid field experience and research programmes. The whole farm is open; there is no closed door or subject taboo. We seek to be as truthful as possible by emphasising our errors and problems.

Internalise and connect: trainees who want to become farmers arrive with dreams and anxieties. This is fully integrated during training. There are spaces made for addressing personal and intimate subjects, sometimes loaded with strong emotions.

Every morning there is a time for physical awakening: gardener's yoga, small massages, stretching, sometimes led by one of the trainees. This unites the participants.

Being friendly: every Thursday, the last evening of the week, a party is organised. A sumptuous dinner is prepared by Fabien, our chef, and by the instructors. Trainees are invited to bring a culinary specialty or a drink of their region. Musical instruments are welcome. These evenings mark the climax of the week, when participants know each other and have formed friendships. Everyone feels better equipped to tackle their project and richer in the support from others. The atmosphere is extraordinary.

Follow-up: trainees can stay in contact with each other as well as with the farm. We strive to always respond to their questions.

Training of trainers: we give enormous importance to this aspect. There is such a mismatch between training needs and the small number of trainers available. We hope that one day soon the territory will be rife with farms like this one.

Several thousand trainees have been trained at one time or another on their journey at Bec Hellouin Farm. They are involved in many projects and initiatives now, in France and beyond.

Perhaps I tend to emphasise only the positive aspects of training. There is a price to pay to get there. Each person on the team gives so much that we end the weeks exhausted (but fed and happy!). For Perrine and me there is no rest possible, neither before nor after. We could have made the choice to live comfortably with reduced agricultural activity with just a few educational gardens. But we absolutely want the farm to remain a real farm and earn a living from the sale of its produce. We must, therefore, continue to manage all aspects of the farm, fill the orders, and open the store, even during training.

The desire to be an open farm and to share everything that can be done is not at all comfortable. We are constantly exposed to scrutiny and the judgement of others. We have taken a few hits, but without hesitation, we can affirm that the adventure was really worth it.

▲ *Thursday evenings, the last evening of the training week, are devoted to a celebration that can continue late into the night. Cécile and Gérard, two solid pillars of our training, here show the local products brought by the participants, before a sumptuous buffet prepared by Fabien and the team. It is traditional that the evening ends in the river... At least that's what I was told, because at this late hour I'm asleep.*

▼ *A certificate of internship completion concludes the training.*

◄ *A discussion in the greenhouse with François Léger, the scientist who launched the research programmes at the farm. The encounter with François was beneficial for us. We learned so much. His intellect and scientific approach made it possible to produce knowledge that gives credibility to a new agricultural model.*

24

GENERATING KNOWLEDGE

The first thing that is missing here [in the West] is this knowledge that comes from practice.

The second thing is the courage to apply it. People are afraid... In our cultures, courage was taught.

Blue Eagle, Cherokee

▼ *Since the launch of our organic permaculture market gardening and economic performance programme, we have welcomed researchers, agronomists and students from many different countries. Scientists are much more convinced of the urgent need for agricultural practices that are respectful of the biosphere, than politicians.*

Developing ecoculture requires moving forward in unknown ways. Like the navigators who come before them, farmers need to draw up their plans from scratch. Lack of scientific benchmarks and techno-economic analysis is glaring and penalises the progress of new farmers. Many aspects of ecoculture need to be documented.

Every farmer can contribute to the bank of knowledge. These contributions can take multiple forms.

Observe, Note, Quantify

When farmers decide to make theirs a farm that generates new knowledge, they are the main beneficiaries. They focus on the finer details and begin keeping records of different parameters like investments, labour time, production levels, and efficiency of crop associations. The data collected allows trainers to be more precise with interns and during training sessions. This information can also feed collective databases.

Participate in a Research Programme

Another option is to become a research centre that is managed by an educational institution. This partnership could be a single farm or several, in a regional, national or international operation.

We were pleasantly surprised to see that programmes conducted in partnership with research organisations on our farm have led to the creation of other programmes, both in France and abroad, and are contributing to the field of interest with researchers affiliated with these organisations. The National Institute of Agronomic Research (INRA), for example, has made several films about our first programme, *Permacultural organic market gardening and economic performance*, that received a high volume of traffic on the internet. These kinds of developments indicate strong public interest and encourage decision-makers to pay attention to these innovative themes.

Following several regional programmes, the first national programme to study micro-farms is about to be launched in France.

► Annual meeting of the farm scientist committee that brings together researchers and practitioners from diverse origins. Some greats of organic farming, such as Philippe Desbrosses, president of the Sylva Institute, Claude and Lydia Bourguignon, Marc Dufumier and Dominique Soltner, have given us their trust. There is no doubt that their moral and scientific backing has helped to ease some of the scepticism that the agricultural world might feel about ecoculture.

What Skills?

To participate in an official programme, a degree in agronomy is not necessary: the only qualities required are rigorous observation, good note-taking, and perseverance.

Often far from the realities on the ground, agronomists are keen on creating close partnerships with farmers; we need each other. The agriculture of tomorrow is being invented and will take place more on farms than in research labs, and according to François Léger, the researcher who guided us in our programmes, the important innovations in agriculture all come from practitioners.

What Protocols?

The first step for a new research project is to find and formulate the major questions. These are research subjects that serve the farming community. The second step is the development of a protocol for collecting relevant, applicable data. This is not necessarily obvious with a subject as diverse and complex as a microfarm. The protocol must be simple enough to be followed daily by people who have no scientific background and are already very busy running a farm. And at the same time, sufficiently complete to be useful to scientists.

It took about a year of trial and error to refine the protocol of each of the studies at Bec Hellouin Farm. Data collection begins before the protocol is completely fixed. It is by moving forward that we note possible improvements.

On the following page you will find tables for past and current research, updated daily by the farming staff.

▼ Here are three researchers that we particularly like, left to right: Pablo Servigne, Gauthier Chapelle and Nicolas Vereecken. They really left an impression on us. The works of Pablo and Gauthier on societal collapse had a profound influence on our vision of the world and on the management of the farm.

BASIC DATA RECORDING SHEET – MARKET GARDENING

DATE:	STAFF MEMBER:

PLOT	CROP		SOIL PREPARATION	PLANTINGS			MAINTENANCE	HARVEST			
	SPECIES	VARIETY	(MINUTES)	SOW (MINUTES)	HARVEST (MINUTES)	QUANTITY	(MINUTES)	TIME (MINUTES)	QUANTITY	UNITS (KILOS, BOXES, PIECES)	CROP EXPIRATION (DATE)

OBSERVATIONS

BASIC DATA RECORDING SHEET – FOREST GARDEN

DATE:	WEATHER/TEMP.:	PLOT:	STAFF MEMBER:

STRATUM	TASK		HARVEST						COMMENTS
FOREST FLOOR, HERB, SHRUB, UNDERSTOREY	HARVESTING, PLANTING, FERTILISING, WEEDING, WATERING, SIZE...	TIME SPENT (MINUTES)	CULTURE HARVESTED	TIME SPENT (MINUTES)	QUANTITY	UNITS (KILOS, BOXES, PIECES)	VALUE IN EUROS	DESTINATION S/P/F	

OBSERVATIONS

Destination of harvest: S = Self-consumption P = Preserving F = Fresh sale

Time spent corresponds to the total time dedicated to an action.

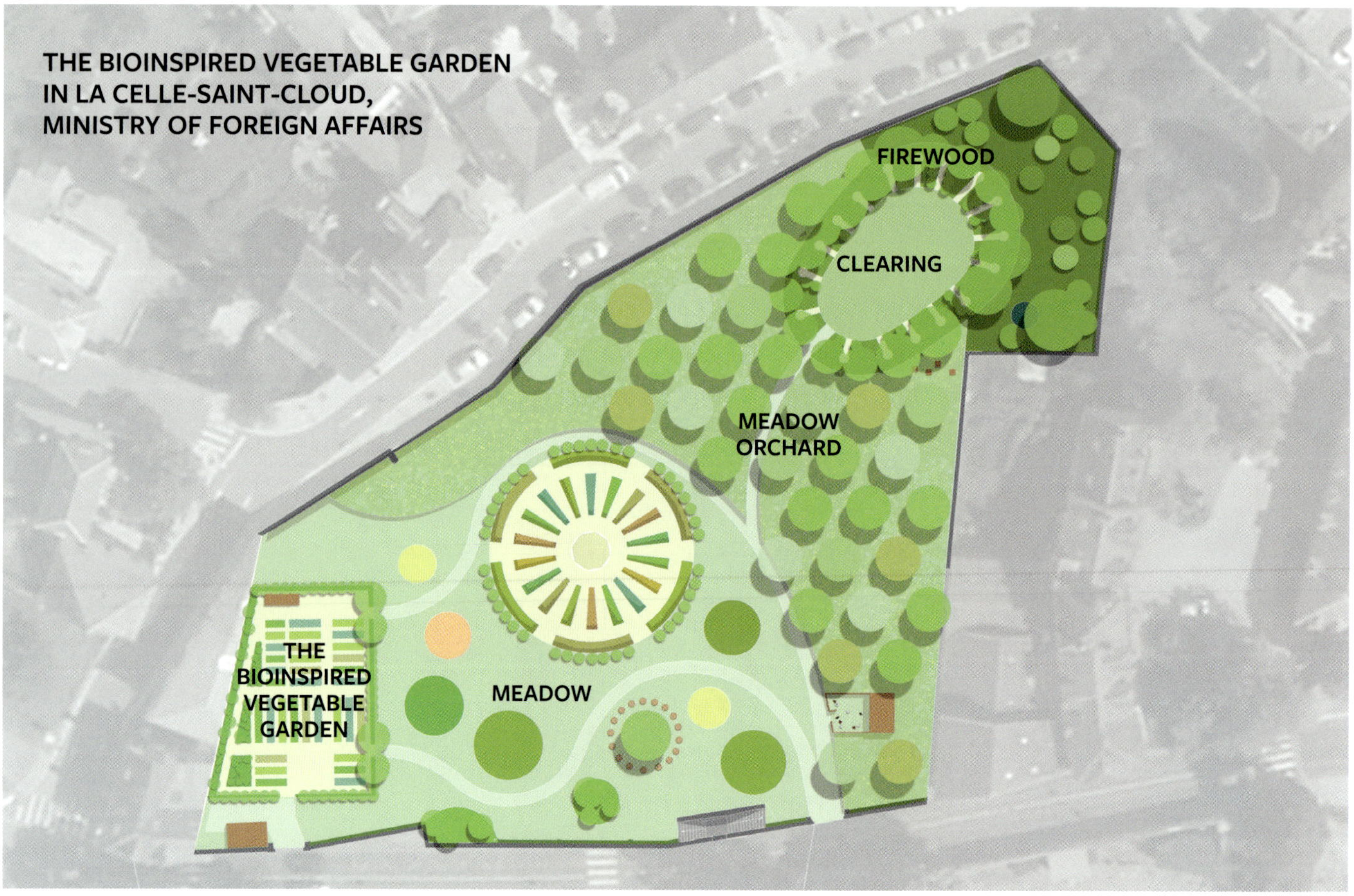

PRELIMINARY WORK
AND DEMOLITION

EXCAVATING

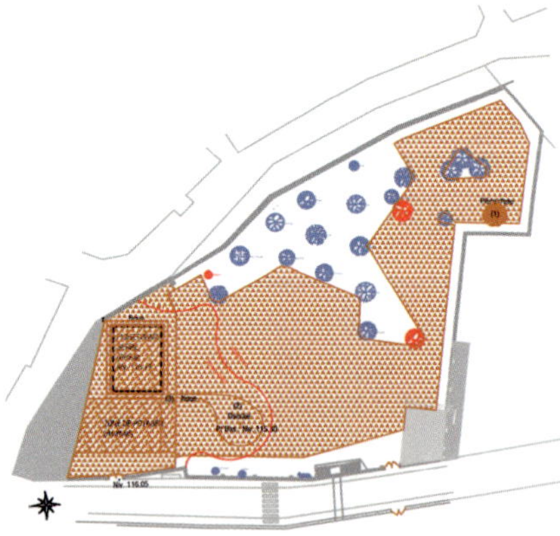

EARTHWORKS

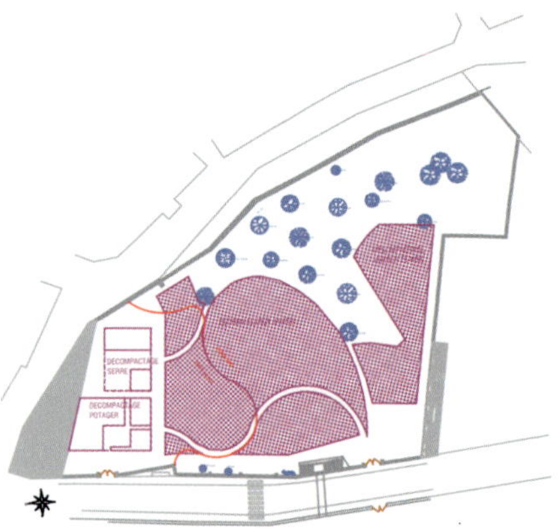

DECOMPACTING SOIL

OUTLINE

PLANTED UP

◂ *For COP 21, the Ministry of Foreign Affairs asked us to design an 8,000 square metre permaculture vegetable garden, intended to be visited by distinguished guests of France, and local residents. The design had to be beautiful and convincing, provide the ability to supply quality vegetables for the minister's table, and be managed by a single gardener. Our experience led us to give a large plot to zones 3, 4 and 5, and to concentrate the biointensive crops in a small perimeter.*

25

DESIGNING A MICROFARM

A farm is nothing like a factory. It is a natural organic microcosm made by the hands of man.

Eliot Coleman[94]

This chapter is of great importance because it integrates almost all the elements covered in this manual. It deals with the harmonious orchestration of all the different aspects of a microfarm.

The term 'orchestration' fits well. We could use the following comparison: a conventional farm growing three crops is like a three-piece orchestra; playing together won't be too complicated. But a microfarm integrating a large number of plant species, animals and activities will be like an orchestra of fifty or a hundred musicians: playing together is a real challenge. The risk of cacophony is great. Reaching harmony will require a lot of design work upstream, then a lot of observation and tweaking.

On our farm, the difficulty of coordinating all the parameters - activities, cultivated areas, work times - was the main source of problems and exhaustion. Most of the time we suffered through the development of the farm without being able to plan or decide. It has only been in the last couple of years that we have had the chance to review the design of the farm in depth. We made several important strategic decisions and remodelled the farm, taking particular care to factor in time and workload, so as not to find ourselves permanently overwhelmed.

▾ *This project gave us a chance to meet the talented landscaper Rémi Algis. We work in collaboration with him on all major projects. Rémi brings his rich experience to create plans and provide guidance for the installations. We believe that being trained in permaculture is not enough to satisfy projects of a certain magnitude, and that we have everything to gain from partnering with a multidisciplinary team that includes landscapers and urban planners, among other types of disciplines.*

The Specifics of a Microfarm

We draw your attention to a very important point which is missing in agricultural training and farming literature because the subject is too new to be treated. We only really became aware of it from our first study!

As you know, nothing in life is straightforward. Each farm has unique strengths and weaknesses. The power of the style of micro-agriculture taught in this manual is its ability to intensify production over a small area. In market gardening, the results obtained show that we can multiply the value of the production almost by a factor of 10 per surface unit. This has been widely explained in these pages. Another major advantage is that the rest of the land is available for other uses. This is a great opportunity to diversify activities and become sustainable. Diversification makes it possible to increase the sources of income and number of jobs created.

Because a microfarm is a complex system, it becomes more difficult to assess workload overall: we will have to take care of the intensive core of the farm, but now also the rest of the territory.

Most project leaders already have difficulty estimating the working time they will need to devote to the main activity. They fail to question the time required by other productions, maintenance time and more. But the days are not extendable. It is important and necessary to estimate the burden of overall work required. If this turns out to be too great, changes need to be made: decrease the area cultivated for market gardening, if this is the main activity; decrease the degree of intensification by reducing the number of succession crops and grow species that require less; get help or even create a new position. In all cases, this modifies both the workload and the revenue, so a reassessment of economic viability is required.

Hence the importance of excellent design. At Bec Hellouin Farm, our new design essentially consisted of a redefinition of the zones, in order to reduce the surface area of those

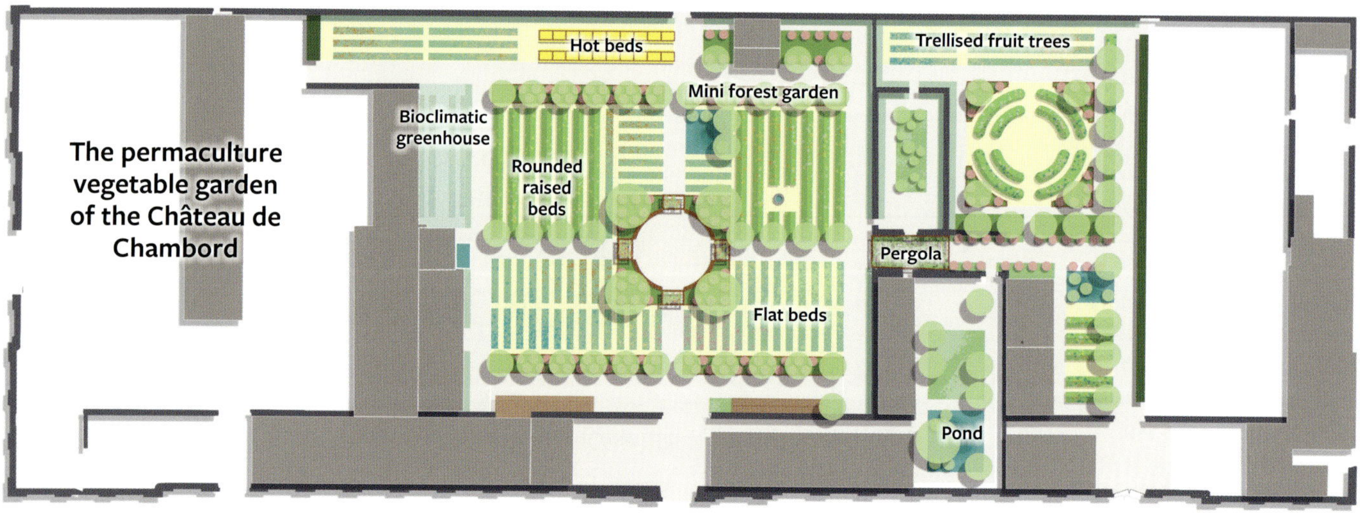

▲ *In 2018, I had the joy of working on the design of a microfarm for the Château de Chambord, with Rémi Algis. Their dynamic management team wanted to reconnect with the agricultural traditions of the region in an innovative way, in sync with contemporary ecological issues. The first phase of the installation is located around the stables of the Maréchal de Saxe. The Domaine de Chambord welcomes more than two million visitors per year, to the different attractions on its grounds including shows, educational workshops, a gift shop and a place of restoration. The microfarm needed to quickly reach economic autonomy, so it was necessary to reconcile agricultural production objectives with tourism and educational objectives. Permaculture design is fully adapted to this type of challenge. The plan integrates the wishes of their design team for a classic layout that is reminiscent of French gardens.*

that require the most labour, which has the effect of reducing the number of working hours. Concretely, some market gardens have been converted into meadow orchards.

An Equation with Four Unknowns

The modelling of a farm depends on a multistudy of factors, but these can be grouped together, divided into four main categories:

- The cultivated area,
- The activities practised,
- The number of hours of work necessary to complete each activity,
- The economics of each activity.

All of these parameters are interdependent, so it is necessary to adjust them in relation to each other. Calculate the surface area cultivated, with working hours and money. Adjustments to the operations must not require more labour than we are able to provide and must generate enough profit to ensure its viability. And, of course, the end goal is that the farm resembles the one you dream of!

In the case of a market garden farm, if the workload is too great, we can reduce the surface area cultivated and dedicate this space to animals. But every square metre not cultivated reduces production. If the other activities carried out on the site allow us to still achieve economic balance, everything is fine. If this is not the case, you will have to change other parameters to achieve this. Perhaps open the farm to paid visits, or organise weekends of training, or host a jam making class.

Designing a microfarm involves playing with all these parameters in order to best match them to your aspirations in terms of quality of life and of income. It is infinitely simpler and less expensive to think about it on paper, before construction! This refers to the need to have indicators reliable for each parameter, which is far from being the case... We will see below what solutions can be tried to compensate for this lack of visibility.

What Indicators for Market Gardening?

Quite frankly, the reference values used to estimate labour hours and potential income, per activity, are very random. These indicators depend on many variables. A few include: climate, soil quality, labour efficiency, organisation, sales outlets... The speed to accomplish the same task can vary by a factor of four, between a slow beginner and a fast and experienced farmer.

For most productions, technical literature gives references and averages that offer a general idea.

For organic market gardening in France, Camille Joyeux carried out a meta-study on a grouping of various completed studies. Her report on technical-economic data in organic market gardening in France is available on our website.[95] It highlights the very great disparity of results according to the techniques used.

For permaculture market gardening, practised according to our method, the Organic Market Gardening study conducted at Bec Hellouin Farm is the only study to offer precise benchmarks to date. Personally, I use these benchmarks in my forecasts. However, it should be kept in mind that this is data collected eight years after the farm started. Depending on your context and your expertise, you will experience better or worse at your farm. During the first few years, it's a safe bet that you won't reach these numbers.

We have already cited these benchmarks in several chapters of this three volume-manual, but their importance justifies presenting them again here. This is the data taken from year three of the study:[96]

- The cultivated area studied was 1,000 square metres of flat permanent beds and rounded beds.
- These 1,000 square metres required 1,600 hours of work over 12 months for all operations related to crops and harvests.
- The value of actual commercial production was 54,600 euros.
- Each square metre generates around 55 euros (all surfaces combined).
- Each square metre under polytunnels (hoop houses) generates a value of around 80 euros.
- Every hour of work in the gardens produces 34 euros.
- We find that adding 50% more than the time spent working in the gardens accounts for tasks related to maintenance, marketing and management. The burden of annual work therefore amounted to 2,400 hours (before the addition of the campagnole which lightened the workload). This volume of hours corresponds to annual full time.
- All work and task time combined result in approximately 23 euros per hour.
- The cost of tools and equipment that made it possible to carry out this production (including polytunnels, irrigation and more) amounted to around 22,000 euros.
- For an equal amount of labour, the smaller the cultivated area, the greater the value produced.

Forest Garden Data

There are no techno-economic benchmarks in available literature for forest gardens. The very first indicators come from the study of our mini forest garden in year one.[97] This data is not of great value because it will take 10-15 years to have enough of a perspective. For the moment, I am basing it on the same value created per hour of work for market gardening, i.e. 34 euros (in year one, our mini forest garden has created a turnover of 37 euros per hour with just the herbaceous layer), and productivity of 15 euros per square metre for an intensive model. This is obviously very random.

Calculate Your Own References

One conclusion is necessary: do not give too much value to the figures of others; consider these to be basic trends. Create your own references, using the data collection grids that we created. They are presented on p.213. One piece of advice: these are for your personal use only; do not try to achieve the level of precision and rigour required within the framework of a scientific research programme, because this would take a lot of time, time that would be added to all your other tasks. If you dedicate 15 minutes a day to recording and analysing data, this should be enough to allow you to have useful reference points. Simplify your statement sheets.

Integrate All Data

Collecting data is a first step; the next is to integrate it. If you are comfortable with computer work, you can create a single document that integrates all the parameters and does the calculations. If you don't know how to use a spreadsheet, you should find a friend who will help.

I made a spreadsheet like this for our farm. It's fun and not very complicated, and above all very practical because if I modify one value, the others are adjusted instantly. For example, if I reduce the area cultivated in market gardening zone 1, turnover and working hours are revised and lowered. I tried not to forget any task, even those that do not generate any specific outcome. I used data from our study, adapting them for areas cultivated in animal traction, for which I have no precise data.

The conceptual tools of permaculture can be applied with effectiveness in all kinds of contexts.

▼ Built during the reign of Francis I, inspired by Leonardo da Vinci, the Chambord estate holds the largest Renaissance castle. Its 500th anniversary was celebrated with splendour in 2019. It is a beautiful place to visit.

▸ *The main problem posed by a microfarm is the multiplicity of tasks to accomplish. If we diversify too much, they risk adding up until they become suffocating. There are the activities that we decide to take on, and others that were not anticipated. Don't underestimate the fact that an innovative microfarm will be influenced by many actors. When you don't know how to say no, you can quickly become overwhelmed and as stressed as you would be living in the heart of a big city!*

Remember, permaculture design is a perpetual evolution. This is normal; there is no problem with that!

Reading the above, a paradox emerges: we must plan, design, anticipate... but we lack reliable benchmarks, especially before installation. This calls for caution. It is like when you make a forecast budget; it is better to reduce the expected revenue and increase the anticipated charges, reducing the risk of unpleasant surprises.

Sailors know to prepare for the worst before starting a long journey and prepare for the worst conditions of a huge storm! Everything on board is designed to withstand extreme conditions. It is likely they will never show up, but being prepared makes all difficulties surmountable. It's the same for your microfarm. During the first few years, I suggest you prepare for difficult conditions and protect your back; that way you will have only pleasant surprises! After three years of operation, you will start to see clearly.

We strongly advise you to be vigilant on the following points. Start with a very small cultivated area and build up gradually. This advice is the golden rule that you can tattoo on your forearm! It applies especially to market gardening in permanent beds. It is better to cultivate a small area and treat it intensely and making more successions, for example. Make four successions in one bed rather than two successions on two beds: the overall workload will be lower due to the fact that the preparation of the soil, the fertilisation, irrigation and especially weeding will benefit four crops rather than two.

In year one, if you apply our method, it would be reasonable to just cultivate 400-500 square metres of permanent beds. If you see that things are going well, increase the intensity of the crops, add a new succession. It's infinitely better than starting on 800-1,000 square metres and finding yourself overwhelmed, with beds covered with weeds that force you to start everything almost from scratch.

For open-field crops with animal traction and arboriculture, the care factor influences productivity less directly than for market gardening on permanent beds. One can intensify care beyond a certain threshold. The number of square metres cultivated or the number of trees and shrubs directly influence the turnover. If you want to become a sylvanier, you will need to plant enough trees and shrubs. A currant bush or an apple tree, for example, will not produce eight times more because you care for them intensely, while a permanent market gardening bed will do just that.

Create one activity after another: this point is addressed in Chapter 15 (p.155) - do not launch three or four workshops at the same time. In year one, focus on your main activity and master it well. Eventually, if you feel like it, add a second activity.

Integrate the Time Factor

Designing a microfarm cannot be an evolving design: even before the first shovel, think of all the activities that you plan to eventually integrate on the site, reserve the necessary space for future installations and consider all aspects of the future farm. But plan for a gradual ramp-up. Define the phases: what you are going to achieve in year one, then in year two, and so on.

◂ *Administrative tasks which are required to manage a farm are considerable; they constitute a real burden for farmers. If we are not vigilant, the time that is spent in front of a computer can become exponential, to the detriment of production activities.*

At the end of your first year, you will be able to make an initial assessment of your organisation, draw lessons from successes and failures. You will be better able to plan your year two... And do not be disconcerted if you have to rework your design in light of what actually happens in the field. Be flexible!

An Example of Design: A Diversified Microfarm of 1 Hectare

We will now present to you two sample models. These farm designs are hypothetical, but I found it more pleasant to present them in the form of stories. I like both, and if I had to start from scratch, this is the type of farm that I would create, especially the second model.

This 1 hectare microfarm integrates all the concepts presented in this manual. It was created by a couple of neo-rural people, Claire and Tom, 10 years ago. They were 40 years old and both had a good first career: Claire was a psychologist and Tom an entrepreneur. They had capital that made it possible to make the initial investments. During the first two years, Claire and Tom had local craftsmen construct the buildings and the dwelling house where they live (happily!) with their children. Their objective: food self-sufficiency, a simple and serene existence, free time while benefiting from sufficient income to provide a good education for their children and be secure during their old age.

Their microfarm allows for diversified activities: intensive manual market gardening on flat beds, market gardening using animal traction, and arboriculture. Animals are present there as well as grain crops; they are not sold but contribute to achieving the objective of food security.

All agricultural production is sold in the farm shop.

The drawing overleaf is an artist's view, however it is approximately to scale. I started from a plot of 100 metres by 100 metres, and couldn't help but add a river. I integrated the lessons learned from the Bec Hellouin Farm. Thus, the store and its parking are located very close to the entrance. The gardens in the intensive core of the farm come next, easily accessed by customers.

The residential house, for its part, is isolated in order to preserve the privacy of its inhabitants.

The drawings and table (pp.222-223) show the zones, an estimate of the working hours that each activity requires and an estimate of the production.

A 1-HECTARE
DIVERSIFIED MICROFARM

1 Edible forest
2 Beehives
3 Meadow
4 Animal traction field
5 Polytunnel
6 Chicken coop
7 Meadow orchard
8 Woodshed-workshop
9 Glasshouse
10 Animal traction field
11 Fruit hedge
12 Stable
13 Granary, storage
14 Shop
15 Tool storage
16 Flat garden beds
17 Animal traction field
18 Animal traction field
19 Meadow orchard
20 Parking
21 Mini forest garden
22 Fruit hedge
23 Mini forest garden
24 Animal traction field
25 Fruit hedge and fodder

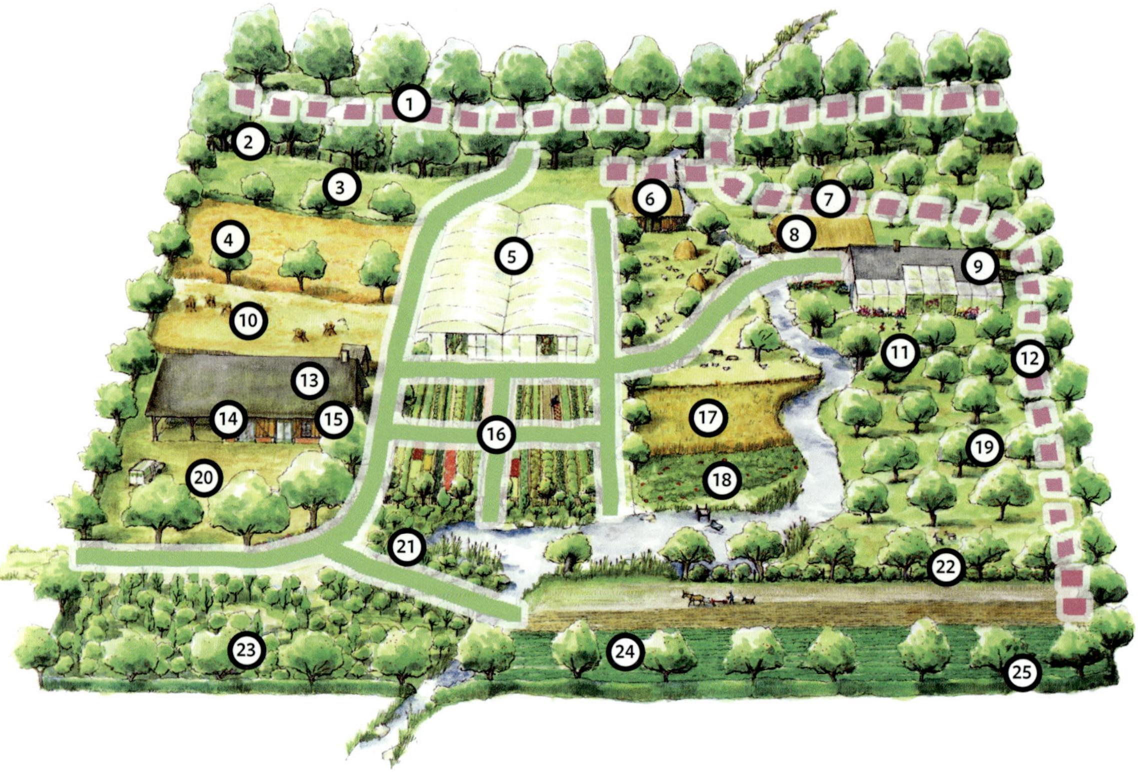

Pathways

Pathways in the edible forest and between fields

ZONE 1

A Polytunnel (600m²)
32 flat beds (230 m²)
B Glasshouse

ZONE 2

C, D Farmyard enclosure and pigs (350m²)

ZONE 3

E Meadow (600m²)
F Meadow orchard (600m²)
G Meadow orchard (840m²)

ZONE 3

H Traction field animal (460m²)
I Traction field animal (254m²)
J Animal traction field (860m²)

ZONE 3

K Mini forest garden (700m²)

ZONE 4

L Edible forest (1,500m²)

MARKET GARDENING ZONE: FLAT BEDS	
Total surface area in market gardening zone 1	920m²
Area actually cultivated	630m²
Area cultivated under polytunnel	400m² + plants
Area cultivated outdoors	230m²
Working time	1,153 h
Turnover – outdoor	230 x 50 = €11,500
Turnover – polytunnel	400 x 80 = €32,000
Turnover – total	43,500 €

MARKET GARDENING ZONE 3: ANIMAL TRACTION	
Grain field	460m² (chickens and self-consumption)
Corn field	154m² (chickens and self-consumption)
Squash field	100m²
Large field	860m²
Turnover – fields	1 or 2 rotations per year × 5 €/m²
Developed area	2,534m²
Turnover	€12,670
Working time	€12,670: €35/h = 362 h

MARKET GARDENING WORKSHOP	
Workload in the gardens	1,153 + 362 = 1,515 h
+ time dedicated to marketing, maintenance, management	50% = 757 hours
Total working time	2,272 hours
Turnover – market gardening	€56,170

FRUITS AND BERRIES	
Edible forest	1,500m² – 26 walnut or chestnut trees
Turnover– estimated	80 €/tree = 2,080 €
Mini forest garden	700 m²
Turnover – estimated	10 €/m² = 7,000 €
Fruit trees	99
Turnover – estimated	60 €/tree = 5,940 €
Fruit hedges	35 linear metres = 35 blackcurrant or gooseberry bushes
Turnover – estimated	32 €/bush = 1,120 €
Turnover – fruits and small fruits	€16,140
Working time	(35 €/h) 461 h
Plus marketing	50% = 691 hours

ANIMALS	
1 donkey, 2 goats, 2 pigs, laying hens, beehives	
Area of meadows + edible forest	2,040m² + 1,500m² = 3,540m²
Fodder hedge	385m²
Estimated working time	360 h/year

IS THE PROJECT VIABLE?	
Total working time	3,523 hours or 1,761 hours/person
Market gardening	2,272 h
Arboriculture	691 h
Animals	360 hours
Various maintenance	200 hours
Turnover – total	72,310 €

This snapshot of a 10-year-old microfarm suggests that Claire and Tom's objectives have been achieved:

- A workload of 1,761 hours yearly per person along with a store of unsold production for food self-sufficiency is well below the industry average. It leaves room to make firewood, partake in hobbies and spend time with children.
- 72,310 euros is enough for the family to live decently on, with expenses paid.

A 'WILD' 1-HECTARE MICROFARM

▲ *A microfarm is a conducive environment for craft creation and artistic endeavours. Natural materials have unlimited value. To be a farmer-artisan seems like a great life, doesn't it?*

Another Design Example: A 'Wild' 1-Hectare Microfarm

Mona and Ian are a young couple who have made some radical life choices to create their microfarm. They met at university where both studied ecology. A two-year journey to South America followed, where the heart of the Amazon gave rise to their project. When they returned, Mona accepted a job as a researcher and Ian learned carpentry. They worked and saved enough to start their 'wild' microfarm.

Mona and Ian seek autonomy and immersion in nature. They started with very modest means and do everything by themselves: planting trees, construction and more. They have little financial needs and lots of free time that they dedicate to their two children, to artisanal and artistic pursuits, and spiritual development.

Both have a strong passion for nature and devote a good part of their time to observing how plants and wild animals occupy the land. The trees have thrived and over the years the land has become a beautiful oasis. This little world apart seems much larger than it actually is and visitors are struck by the ambience that emerges from it. It seems that Mona and Ian have succeeded in recreating a kind of tiny Amazon. Their microfarm is ecologically very efficient.

Their land is 100 metres by 100 metres. The ponds were hand dug and there is no river.

Mona and Ian are sylvans. They planted an edible forest with 10 metre-wide strips of forest garden in such a way that light penetrates and allows maximum productivity. These forest garden corridors surround three clearings dedicated to animals, vegetables and grains. The crops and animals are not sold, but rather enable the family to live self-sufficiently. The family is vegetarian and makes do with milk from goats and eggs from chickens. The goats contribute to the maintenance of the undergrowth. The couple makes a living by selling forest garden products at a local market. Their expertise in botany allows them to also sell wild plants to restaurants.

Ian and Mona offer training in medicinal plants. They plan to launch a new activity that will allow them to acquire adjoining land that they will reserve for wildlife.

In their quest for autonomy, the couple chose not to connect with water and electricity networks. They installed solar panels and a small self-built wind turbine, a rainwater collection system and filters to purify pond water.

A solar dryer and oven are used to dry plants and fruits, and even make jams.

Mona and Ian are passionate about crafts. They use the biological resources of their small territory, like coppiced willow and chestnut trees to make baskets. Ian made a manual wood lathe, with a spring made with a bent hazel branch. They create pottery with clay extracted from their land, which they use as containers for their dried aromatic plants.

Mona studies the feasibility of weaving nettle fibres. Their handicrafts are sold at the market.

► *Our daughters sometimes create small imaginary landscapes, like this leprechaun shelter by the river. A microfarm is like an adult daydream – we create a world that suits us. It is not a rejection of the real world, since a farm is very real, but rather the refusal to accept a degraded and ugly setting that does not meet our aspirations. Our reveries contribute to the charms of Mother Earth.*

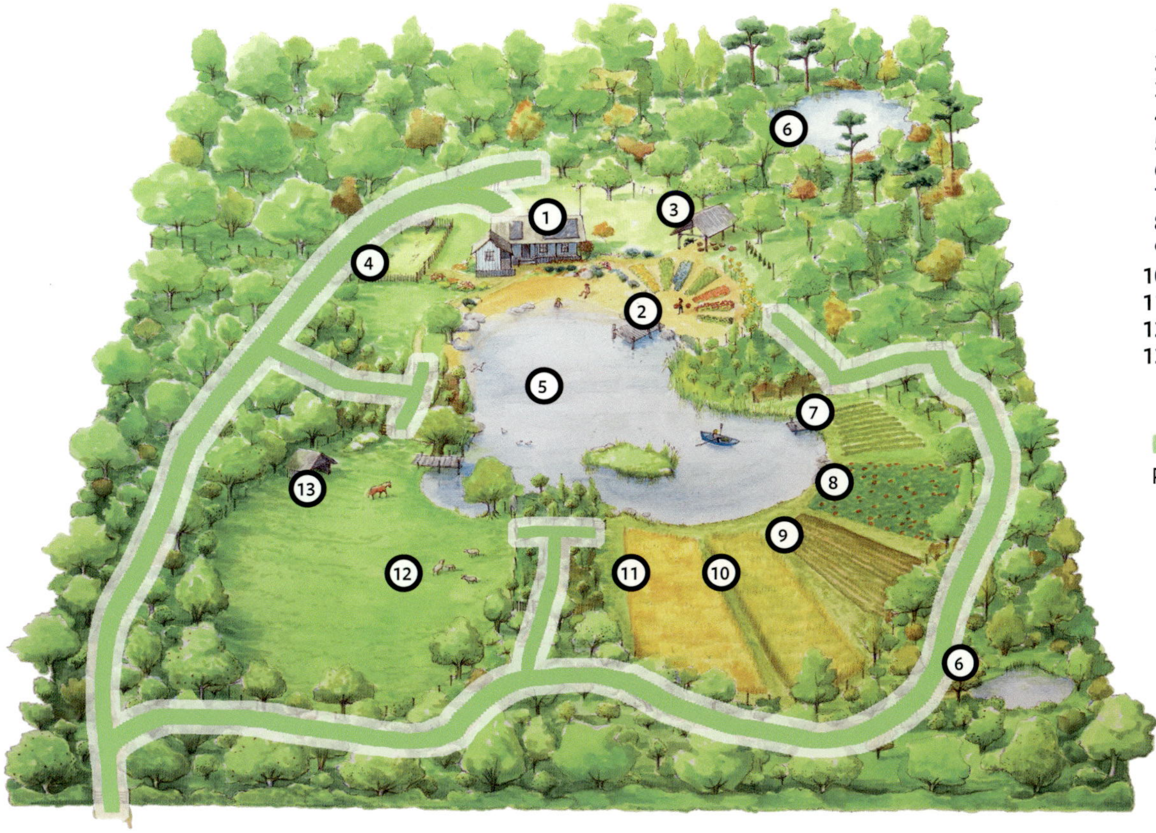

1 House
2 Edible flower garden
3 Plant drying
4 Chicken coop
5 Lake
6 Pond
7 Potatoes
8 Squash
9 Various vegetables
10 Corn
11 Wheat
12 Sheep
13 Pony

Pathways

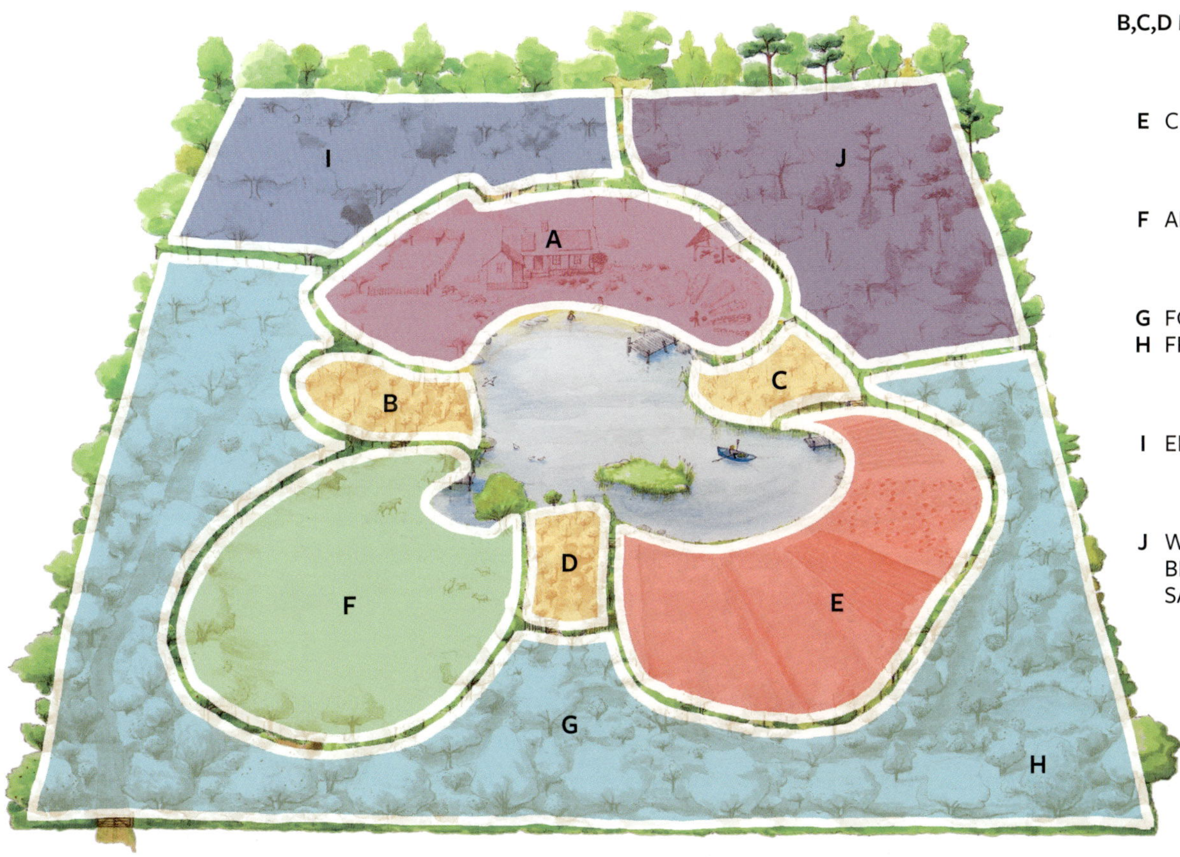

ZONE 1

A HOUSE CLEARING

ZONE 2

B,C,D MINI FOREST GARDEN

ZONE 3

E CROP CLEARING

ZONE 3

F ANIMAL CLEARING

ZONE 3

G FOREST GARDEN
H FRUITING SHRUBS

ZONE 4

I EDIBLE FOREST

ZONE 5

J WILD FOREST BIODIVERSITY SANCTUARY

Villa de la vie

◀ *The big concern, for Perrine and myself, is being available for our four girls. We see so many people missing time with family for the causes they serve. Stress and overwork led us both to the hospital. Fortunately, a farm is a very beautiful place to grow up. The learning opportunities are numerous and varied. The companionship of animals is a powerful balancing factor for children. And, unlike those who work outside the home, we are here for every meal, and for each bedtime.*

26

ADVICE TO FUTURE MARKET GARDENERS

Time is running out. Let us therefore know not to hurry.

A wise man, quoted by Lanza del Vasto[98]

It appears that we are coming to the end of this manual. I admit that I am not upset about returning full time to the gardens – it is the month of May and it is so beautiful outside – but, before leaving you, here is a summary of the advice given to future market gardeners over the course of the various chapters.

▼ *Run with a smile... with the hope of slowing down one day! We really hope that this manual will help you avoid at least some of the countless mistakes we made, and make your life easier. We put everything we had to share into it. The adventure is now in your hands!*

Before the Setting UP

Train Yourself, Test Yourself

If you do not come from an agricultural background, your training will require several years. Immerse yourself in the environment, test yourself in a real situation, in all seasons, and dialogue with market gardeners who practise different forms of market gardening.

Create an Excellent Design

Drawing and designing on a sheet of paper or a computer is cheaper than doing it for real on a plot of land. Dedicate the time needed for your design. Document every aspect. This will bring great savings in labour, time and money.

Watch your back. During the first three years you will be faced with many struggles. This will be your truth test. Plan for what you will live on while waiting for your farm to become profitable.

Don't forget to be happy though. The adventure is exciting!

Create your Network

When you have acquired your land, take time to build connections with local stakeholders. Meet the mayor, agricultural institutions and associations, local farmers, your neighbours and more. Get to know them and humbly introduce your project, both honestly and sincerely.

Do Market Research, Identify your Customers

You should not put your farm into production without knowing what types of products are in demand or to whom you will sell them. Talk to your future customers as much as possible.

The First Years

Adapt your Investments to your Starting Capital

Do not take on investments measured against your ability to self-finance: they would hinder the economics of your farm. Reduce expenses and focus on what pays off. If you have almost

no initial capital, make do with a campagnole and some manual tools at start-up, and begin production with these things. Buy equipment second hand. Avoid borrowing.

▼ *I have long feared that pursuing my dreams, whether with* Fleur de Lampaul *or the farm, was a selfish endeavour. It was only recently that I realised that living according to one's heart can also brighten the lives of others. Our upbringing often carries a burden of fear and guilt that forces us to stay in line. Building the world of tomorrow can't be done with submissiveness and guilt. Let's dare to deploy our wings and fully embrace the adventures of our lives!*

Start Small and Take Meticulous Care in your Work

Don't fall into the trap of believing that the bigger your gardens are, the more you will earn. This may be true with a tractor, but with a manual biointensive approach to market gardening, the smaller the cultivated area, the more your production will grow and your profit margin rise. Look for value in everything that can be grown, including wildflowers and plants.

Gain Speed Gradually

Better to gradually increase growing space when you feel able to care for a larger surface area, rather than abandoning gardens that we cannot cultivate. You need to create excellent market gardening land and this is much easier to achieve on a very small area. Also think about weeding, which can quickly become time-consuming on a large scale, especially in the first years.

The same goes for diversification. Take it step by step; launch a new endeavour only if you master what you started and still have free time and desire.

Work Quickly and Effectively

One of the paradoxes of market gardening is the need to work both quickly and well. The workload is such that, if you are slow, you will not complete it. But if you botch it, the job will only grow. Being both fast and careful is not easy and requires proper training.

Have Excellent Tools and Maintain them Perfectly

If you choose to work entirely by hand, treat yourself to the (small) luxury of having the best possible tools. Put them in your hand, sharpen them, grease them. You will gain a speedy execution and the additional cost compared to mediocre tools will quickly be written off. Not to mention the joy of working with such reliable companions.

Do Not Neglect any Task

During the first years, you lay the foundations of your farm. No building will be sound if there are holes left in the foundations.

In season, market gardening often creates a sense of urgency; we feel we are behind with our plans, with the impression that we will never finish. The temptation to neglect weeding while telling ourselves that we will come back later can be common. It only takes a few weeks to find yourself with a farm in chaos and completely grassed over. This will only increase anxiety and sleep loss.

Take the time to do things right. A baby weed will always become a monster at the adult stage and take 100 times more effort to remove. Add the bed to your 'finished' list when you know it is definitely weed-free. This will then leave you with three to four weeks of tranquillity, without having to return. Weed regularly 'while it's clean'.

Be Well Organised

When you work, be observant and ask yourself about the best way to progress. Try to save a few minutes for each operation; at the end of the week it will make a big difference. This can happen by improving the design – for example bring the tool shed closer to the garden or create a new path, or by choosing a tool that is better adapted to the job. Be organised.

Try to consolidate the harvest into one or two days a week, no more. Plan a time for sowing each week.

Prepare your year carefully in the autumn. Create tools to help work things out, for example an annual seedling plan, annotated and improved from year to year. Keep a weekly calendar of events based on previous years.

Install a large whiteboard in your shed or in your workspace. At the beginning of each week, note all the tasks to be carried out, who does what every day, and check each item off when accomplished.

Keep your Farm Tidy

Nothing is more demoralising than a farm where many things are neglected due to lack of maintenance. Don't lose your footing; do what needs to be done gradually. Don't postpone until tomorrow what you can do today.

Tidy up as you go. Don't leave the farm in the evening without it being impeccable, so you will be happy to return the next day.

We often tend to favour tasks 'that pay', over maintenance and the beautification of the farm. But a tidy, flowering, welcoming farm will attract customers and better collaborators, and you will have much better morale and more energy for work. A beautiful farm is more profitable.

Don't Count On Unpaid Interns to Take On a Workload

It is common to find that a poor design, with cultivated areas that are too large and/or sloppy organisation, plunges farmers into work overload. The temptation is great then to delegate part of the work to interns. But beginner volunteers will not necessarily make your life easier; don't rely on them to do your job in your place. Your farm must be well adjusted to its permanent workforce. If you take on trainees, it must be only for training purposes. Do not hesitate, on the other hand, to organise collective projects, for example for the planting of a forest garden. Be sure that this joint work is friendly and educational.

Do Not Neglect Accounting and Administration

Most of us prefer to be in the gardens rather than behind a computer. But accounting is like weeding – when we neglect it the bills pile up, and the less we want to get started, adding more time to the task.

Ideally, your farm should be clear and well outfitted from A to Z. The ideal does not exist, but be certain that starting small and doing things well, step by step, will make you considerably gain serenity and joy of living.

What kind of world will we leave to our children?

◄ *One hundred and eighty million children work and are deprived of schooling, most often young girls. Nearly 80% of the 830 million people who are currently suffering from hunger are farmers. Families driven from their lands through misery end up in the slums of megacities, where they are subjected to violence, alcohol abuse and prostitution. Too much poverty, like too much wealth, is an obstacle to happiness. Ecoculture has a solution for these problems. This bioinspired approach does not need capital, sophisticated tools, or fossil fuels. It could enable communities of vulnerable farmers in the southern hemisphere to live better from their profession, without generating the addictions of the Western world. Educational programmes in farming villages would make it possible to increase the effectiveness of agricultural practices and restore biotopes.*

27

DEVELOP ECOCULTURE IN ALL PLACES

Science can shed light on the dimensions of man's kingdom in the cosmos. But it will take a political community of men who are conscious of the strength of their reasoning, the weight of their word, the seriousness of their actions to choose freely, the austerity that will guarantee their own vitality.

Ivan Illich[99]

As the writing of this manual began, American and Chinese publishers told us of their desire to publish it in the English-speaking world and in Asia. During the five years writing this manual, we have kept in mind that we are addressing farmers living in all latitudes – it's a bit terrifying if we didn't trust the creativity of our readers and their ability to invent solutions adapted to their local context.

From the Local to the Universal

How does the experience of a small Normandy farm capture the interest of farmers from both hemispheres? Duplicating Bec Hellouin Farm would have absolutely no value, especially since, as we have repeatedly emphasised, our farm is very far from completion.

However, the experience developed at Bec Hellouin Farm can inspire others in all climates, just as we are inspired by practices developed in California, on the border of Canada, Japan, Amazon, Australia, Korea, Indonesia and more. **It is nature that inspires us: the great principles of bioinspired agriculture presented in volume one are universal.** They can be implemented as well by a small farmer in Bangladesh or Kenya as by a farmer in the Middle East or France, with the necessary adaptations.

The intention of this manual is to offer you great ideas and explain how they have been implemented on our farm in Normandy, France, as an illustration of what can be done. Make no mistake: **it's about adapting the principles to your specific context, and not imitating our farm.**

Although we rarely leave Bec Hellouin, we had the opportunity to read professional literature and scientific publications related to our interests. We have also welcomed agronomists and farmers from different continents and carried out consulting missions in both hemispheres. We benefit from the valuable support of a scientific committee composed of researchers with international experience. We hope to have avoided the trap of locking us into our own particular situation. Bec Hellouin Farm is not the centre of the world. We constantly seek to reconcile an anchoring in the very local with an openness to the universal.

Most agricultural textbooks provide standards, like plant your carrots and leeks a certain number of centimetres apart from each other, for example. The agriculture world loves standards! But standards can mislead novice market gardeners, who feel secure with these precise numerical markers, but who are not sufficiently aware of the fact that microclimates and other factors influence crops. What is suitable in one situation is not necessarily the same in another. Does copying planting distances prescribed for California or Normandy really make sense in your local context?

To avoid falling into this trap, when we quote technical data related to our practices at Bec Hellouin, we insist that they are only shared for illustrative purposes. The heart of the message contained in these pages lies in these 'aids to the design of a bioinspired system', these main principles, these 'metarules' proposed throughout the chapters, which will allow everyone to design their own garden.

▲ *Having working animals is a luxury in countries of the southern hemisphere, where one billion farmers work entirely by hand.*

▼ *Increasing amounts of desertification makes vast areas uninhabitable, spawning desperate migrations. The Western world is reluctant to welcome migrants from the south, while the northern hemisphere is mostly responsible for the climate change that drove them from their lands. We are just starting to realise that we must reconcile questions of social justice and ecology.*

Appropriation is a basic tenet of permaculture.

Designing your own unique system requires a lot of investment and a lot of time. But the results will be incomparably superior to those of farmers who are content to simply apply directives coming from elsewhere.

Intelligence of the Living World

Humanity must today repair what it has degraded by neglecting the great laws of life. We must give the earth a fertile coat of humus and luxuriant vegetation, clean the waters and stabilise the climate. Every gardener and farmer can participate in this great effort to enable future generations to live in peace in a balanced world.

As the writing of this manual progressed, we felt increasingly connected to the billion farmers who work with bare hands and simple tools.[100] Nearly 85% of farms in the world are less than 2 hectares.[101] This is a good indication that a manual approach on a small surface area, as described in this book, is suitable for all kinds of contexts.

At the same time we were constantly surrounded by people who, like us, grew up in cities and experienced the abundance of modern life, and who nevertheless aspired to come back to the land. Farmers of the south and north need each other to save the planet. Each of us holds a part of the solutions.

We can all seek to understand the great principles of the living world and meditate on them. We are enriched by the experience of others and others are enriched from our experience. We need agronomists and researchers as well as labourers of the Earth with calloused hands; there is no need to have studied at universities or read books to gain an intimate understanding of nature.

The beauty and nobility of the farming profession resides in the capacity to work with hands, head and heart within the great movement of life on Earth that has been morphing nature towards richer, more complex forms for over four billion years now.

Expertise of Farming Communities

Every rural community in the world has women and men who are sensitive to nature and can play a leading role in the development of new bioinspired agricultural practices. These explorers of the future can draw on the treasures of ancestral knowledge and knowledge of the local environment found in each community.

One might think that collective knowledge about nature was erased when local communities were harmed by globalisation. Many farmers were driven from their lands by poverty and ended up in cities, but experience shows that there are invisible, secret transmissions, shared from mother to daughter, that can traverse centuries of oppression and suffering. When the context becomes favourable again, this knowledge can be rejuvenated, like seeds that wait underground for rain with infinite patience.

Our manual will never replace these treasures of knowledge and experience, or the expertise of bare-handed farmers. At best it can reveal, and draw attention to the immense value of this vernacular knowledge and put things into perspective, thanks to experience that has given good results in other contexts. To develop ecoculture everywhere, we must be able to observe and act with our land while opening ourselves to the whole. Let's share our visions and our intuitions!

What will save humanity is the alliance of local views and universal knowledge that today we are, for the very first time, able to share from one community to another, and from one continent to another, thanks to new means of communication.

Adapt the Main Principles at Each Locality

It is therefore not up to us in Normandy, France, to tell others how they can adapt the principles proposed in these pages to their local context. It is up to you to develop your practices by nourishing them with contemporary knowledge on nature and ecoculture.

This will help you observe your land in a deeper manner, and understand its value before the appearance of industrial agriculture. To seek outside experts and people who have had successful experiences in contexts close to yours, to develop your own hypotheses, to test them in the field and to improve them. I guarantee you one thing if you do this: you will not experience another second of boredom until the end of your days!

Share your discoveries as widely as possible. We thank you.

▲ *Rare vegetation adds a bit of green to the sand dunes and stony ergs of the Banc d'Arguin, west of the Sahara.*

From Agroecology to Ecoculture

Agroecology has been spreading around the world, slowly but surely, for several decades now. Ecoculture is its little cousin, its close relative – just younger, the next generation. All of agroecology can be integrated into ecoculture. The same goes for biodynamics. Ecoculture imitates natural ecosystems, thanks to the concepts of permaculture. Those who practise agroecology and biodynamics benefit from all of that acquired knowledge and learn techniques that will improve their approach to growing.

Avenues for Reflection on the Implementation of Ecoculture in Different Contexts

Although we are not really qualified to give advice to farmers living in other latitudes, allow me to add a few thoughts on environments that are particularly interesting. Just a few basic ideas. If they appeal to you, think about how you could apply them to your farm. We are very interested in reader feedback.

In a Desert Environment

Arable land is becoming desertified, impoverishing already vulnerable populations. Bringing life back to regions with advanced desertification is a colossal project. Many NGOs are employed, often successfully, using the concepts of agroecology such as Terre et Humanisme or CARI (Central America Regional Initiative).

The following reflections concern poor communities acting locally, within their means, to recreate soil fertility by recycling their own organic waste.

I was made aware of desertification during two long stays on the Banc d'Arguin, in Mauritania, during our stopovers in the Cape Verde archipelago. There, drought killed thousands of people during the 20th century. This was also apparent in visits to Australia and the Red Sea.

The Banc d'Arguin is where the Sahara flows into the Atlantic. There, sand mixes with saltwater, and after several decades of intense drought in the entire Sahel region, there remains only one tree and one well for hundreds of kilometres around. I was able to witness the harsh existence of the Imraguen fishermen, descendants of slaves who live on the immense deserted beach that borders this part of the African continent. I spent several months with only two and a half litres of fresh water per day, delivered by boat from Nouakchott. I wondered if it would be possible to create a garden in this desert environment. After all, it was green in the Neolithic age, and archaeological evidence of a once large population was still present. Could humans fix what they have done to contribute to this desertification over the millennia?

The conditions on the Banc d'Arguin are extreme, yet organic waste is not valued

▲ *The Imraguens survive on marine resources. Fishing products are shared among all. The solidarity that reigns within the villages is impressive.*

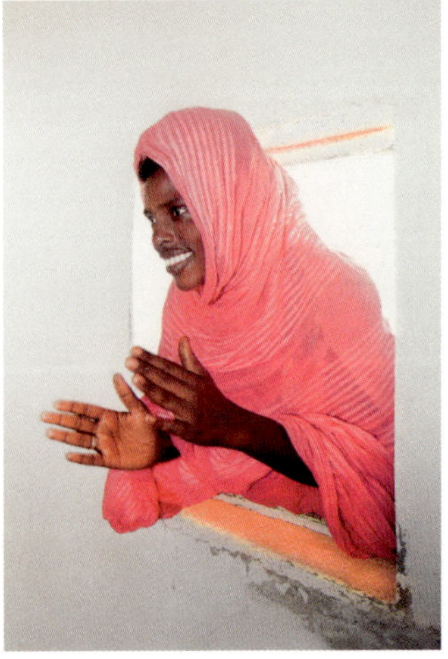

▲ *The main obstacle to food crop development in arid environments is perhaps more cultural than climatic. Education is essential. Women are often the drivers of human development in their community.*

– something I was also able to observe in other desert regions of the globe – all simply because it is not in the culture of local populations. The residents are not aware of the value of their organic waste. In Mauritania as in Cape Verde and in other regions of the world, local populations are highly dependent on resources coming from the outside – water and food in particular, resources often emanating from international solidarity. But the waste from these resources is not valued. This waste is counterproductive, especially since, if we project ourselves into the future, nothing indicates that it will be possible to maintain these transfers of resources in a post-oil world. It would be wise for indigenous populations to benefit from the current situation and enhance everything possible to gradually become less dependent on external aid. NGOs that support them now could play a driving role in this process.

If the remains of the fish, the human excrement, and the algae that wash up on the beaches, along with the ashes of their fires, were concentrated in the same place, would it be possible to transform sand into a substrate where plants could grow?

The permacultural approach encourages us to think about the fight against desertification at the smallest level, that of the individual and the family. Observations of nature teach us that nothing is too small, too trivial to be overlooked. Giant trees from the Amazon know it well – they who do not hesitate to send out roots to tap into the small clumps of organic matter from epiphytic plants that attach themselves to their branches. It is this strategy of cooperation and recycling pushed to the extreme that makes it possible to create bioabundance from scarcity.

If nature can green up isolated sandbanks in the open sea, like Sandy Cay, would it be possible to create islands of fertility in the desert? Observation of the way that vegetation has colonised this sandbar in successive stages has inspired what follows. Even those who live in the most desertified regions of the world drink, eat and produce organic waste – in small quantities, of course, but these by-products of human metabolism are resources where nothing grows. Rather than being dispersed around habitats, they could be used to initiate an upward spiral of fertility.

Urine and grey water, humanure and organic waste could be used to support the growth of a few trees planted very close to the habitat. There are several species of trees, such as acacias, capable of reaching deep water very quickly after the first stages of their growth.

Humanure could be mixed with fish waste, and urine with wood ash to become fertiliser. Seaweed can be used as mulch. The first trees could be planted in a circle to surround a very small clearing of sand that would gradually become a garden. The garden would benefit from the shade of the trees and being sheltered from the winds. Once the trees are mature enough, the conditions would be conducive to implementing a second generation of plants, shrubs and perennials, bushes that survive in the desert and even prickly pears? They would grow in the shelter provided by the first trees, in a ring around the clearing. This second stratum would help improve bioclimatic conditions for the future and become a source of

▶ *In the countries of the south, the implementation of ecoculture consists of synergising the best of traditional agricultural practices and the best advances of contemporary scientists.*

biomass to accelerate the creation of soil. Urine, faeces, kitchen scraps and all organic waste from residents would be placed at the foot of these plants or composted.

In a third stage, the garden clearing could receive human waste as mulch, for the vegetation that surrounds it. Its soil would gradually become enriched in organic matter that would retain water from irrigation dispensed sparingly. This would be a beginning step for food crops.

Over the years, the conditions of this small garden would improve and it would become possible to create a second, then a third. If each family from a village did the same, in the space of 10 years a green belt could be created around the village homes. And if other villages of the region began to imitate them...

The obstacle to this proposition seems more cultural than ecological. Educational programmes and technical support would initially be necessary. A demonstration garden and training could show what the inhabitants of desert regions cannot even imagine.

Our proposal is based on the valuing of waste from a family or a village. At the scale of a small territory, agroecology has developed simple and effective systems to capture and store water when rain occurs. It also helps prevent erosion. Simple technologies, such as solar desalinators and devices to capture water in the air, are being developed. Their implementation in a context like that of the Banc d'Arguin would probably cost less than bringing water by boat, year after year. All of these approaches can be synergised.

With small drops of water we make a river. Everyone can act in their own capacity. Every human being is powerful.

A Garden in the Tropics

We welcomed and trained representatives from an NGO in South Kivu, a once luxuriant region of the Democratic Republic of Congo, that had been ravaged by war and violence. This NGO does remarkable work to care for and support women who are victims of violence and genital mutilation – atrocities used there as weapons of war... The mining resources are plundered and the farmers in this insecure climate survive as best as they can without thinking about the durability of their practices that lead to deforestation and rapid land erosion.

It is by considering possible adaptations to the land that was made available to help these women, that we imagined the concept of a garden clearing (later adapted at Bec Hellouin). The idea started from two traditional models that are well suited to tropical environments, and adjusted thanks to the principles of permaculture.

▲ *The tropics have strong potential for productivity, but are too often ruined by social disorder and unjust treatment with globalisation.*

The first source of inspiration is slash and burn that I observed among Native Americans of the Amazon. These temporary garden clearings are opened with iron and fire, and cultivated for a few years, then returned to forest. In the context of Kivu, the forest was destroyed and it involved restoring forest and food gardens. The second source of inspiration is the Creole gardens that allowed former slaves of the Caribbean to regain their autonomy.

Future generations will live in a pacified world if we can solve the intimately linked issues of social justice ecology.

Each garden clearing is designed as an autonomous entity, enabling it to meet the needs of a family and generate a surplus of sellable products at local markets. It's a small world in itself.

One garden-clearing feature is to grow crops safely hidden from view, behind a dense curtain of vegetation. Each plot is surrounded by a strip of forest garden including a few large trees, like mango trees. The wealth of edible plants among tropical flora creates a stratum of great diversity. The traditional diet of tropical Africa incorporates a large quantity of leaves, increasing the potential nutrition of a forest garden.

The clearing in the heart of the forest garden is dedicated to tubers and vegetables. We suggested that it be divided into four sectors to facilitate rotations, a practice abandoned by local farmers that led to a proliferation of pests. Permanently mulched beds and plant associations promote good soil health, often difficult to attain in tropical environments where pests are active all year round – but achievable despite what promoters of pesticides say.

▸ *Subsistence agriculture that once provided communities with food security for farmers in the southern hemisphere was undermined by cash crops intended for export to rich countries, using the unsustainable practices of Western agriculture, such as ploughing, pesticides and GMOs. The excess produce is sold at local markets as subsidised surplus from northern countries, at prices often lower than the prices of local food producers.*

In order to promote the autonomy of the site, it includes shelter or habitat. Rainwater falling on roofs is recovered. Dry toilets recycle human waste.

Small livestock – chickens, pigs, goats – can be installed in the forest garden. The animals are fed with by-products from the plot and composted droppings. Plants for herbal teas are planted, notably bamboo (one of the causes of local deforestation is cutting down young trees to build scaffolding in town).

Medicinal plants can be grown, notably *Artemisia annua*, recently discovered to have antimalarial properties.[102]

Each plot would be large enough to support a family and produce marketable surpluses. To our knowledge, this project was not implemented. It is never too late to test it!

Coral Islands

The coral islands are pure wonders of nature. They are also extremely vulnerable, notably the atolls and low-lying islands, such as the Bahamas and Turks and Caicos. It is likely that most of these jewels will disappear under the rise of waters linked to climate chaos.

When I was a sailor, I spent years in these coral archipelagos and studied them with biologists. At the time, my focus was turned towards the sea. In two decades, I witnessed the accelerated disappearance of coral reefs due to overfishing, the rise of tourism and warming waters.

I loved the Bahamas archipelago, whose waters are the most beautiful in the world. I had the opportunity to return with my family to a natural park in the Bahamas that we studied with my school ship *Fleur de Lampaul* – bad weather on that trip prevented us from visiting Sandy Cay, the islet mentioned in volume one. Nearly 20 years after my last visit, the degradation of the environment was striking from all perspectives. We spoke with local farmers and others on the various islands. On Man-o-War Cay, we visited a community that has remained attached to its values and collected testimonies of older residents. Life was poor and hardworking in these very isolated places, with the only exchanges with the outside world done with locally built sailboats. The islanders lived from the sea and the land; a bit of agriculture, and a little fishing and craftsmanship, allowed them to achieve relative autonomy.

The contrast is strong compared to what these enchanting islands have become, following the tourist invasion. Everywhere the *American way of life* has transformed Bahamian society profoundly. The islands have become the suburbs of wealthy Americans. They produce almost nothing, and almost everything is imported. We were surprised to discover a massive supermarket on the island of Eleuthera. The archipelago is permanently crisscrossed by countless ships of all sizes, often with oversized engines. Just off their flight, tourists smear themselves with sunscreen and populate these formerly isolated islands. The lost islets are now covered with multi-million-dollar villas.

The natural sanctuary has been violated. All or almost all is tainted by money.

As permaculture farmers, this development represents something surreal. If deprived of oil, and hence tourism, Bahamian society would have zero resilience. If water and food could no longer be imported, it would not even be possible to return to the frugal lifestyles that allowed local communities to survive before the explosion of modernity. Marine resources have been plundered by overfishing and poaching, and battered reefs. As for arable land found on some large islands, they were exhausted by pineapple monoculture crops destined for export. Furthermore, the size of the current population is incommensurable with what it once was and the habits of consumption have evolved so much that we wonder what would become of these island communities if an oil shock suddenly isolated them. It is highly likely that they would perish from hunger and thirst.

In 1492, Christopher Columbus first set foot on the other side of the Atlantic, landing on the Bahamas. He was warmly welcomed by the Arawak Native Americans, a peaceful people who lived naked and travelled by canoe among the thousands of islands in the archipelago. A single generation was enough for the 'discoverers' to exterminate every last one of them. The Arawaks did not survive enslavement by the white Europeans.

Five centuries later came the destruction of this natural paradise. From ethnocide to ecocide, is our civilisation condemned to destroy everything in its path?

What is true for the Bahamas is also true for most coral archipelagos. A coral island is a limestone and sandstone outcrop on the surface of the ocean. Fresh water resources only come from rain. The thin humus layer is extremely vulnerable. For local populations, most of the resources came from the sea, before the destruction of the coral reefs deprived them of this.

These sublime islands concentrate multiple forms of vulnerability. It is eminently urgent to think about their resistance and take advantage of current resources to adapt them to the conditions of tomorrow. But who cares? Everyone seems more interested in capturing the windfall of dollars coming from yachts, charters, nightclubs, and casinos.

Inhabit the Earth as an Assassin or as a Poet?

What an extraordinary project it would be to restore an island and its surroundings, both on land and under the sea, with a vision of giving it back its autonomy! Permaculture logic is particularly well suited to face this challenge. It would be necessary to create a design integrating all aspects: water, land, food, energy, materials, living together, relationships with the outside world, for each island. For the young people of the Bahamas, Turks and Caicos, Polynesia Micronesia, the Maldives and elsewhere, creating holistic and integrated management plans of the islands and their surrounding environment would be an exciting adventure from an ecological, scientific, technical and human perspective.

On an island, the lessons learned from observing nature are of absolute relevance: complete the cycles, generate no waste, value every resource, even the tiniest... And create soil everywhere. Humus is life!

Starting today, let's think about adapting our societies to the conditions of a post-oil world. The worksite is of an unprecedented scale. This must not intimidate us. Let's mobilise all resources, of heart, intelligence and creativity, and let's make it a fantastic shared adventure! The transition from a materialistic society that blindly wastes limited resources, towards a sustainable society is a source of multiple opportunities. It can create work for everyone and give meaning to our lives.

Let's use the last barrels of oil wisely to restore local resilience to every part of the globe. Let's ensure the safety of generations to come. Because the Earth is an island, like Sandy Cay, an island lost in an open ocean, and it is we who can lead her towards death or towards life.

We have destroyed so much; it is time to move forward together on a path of repair and healing. Our choices and actions today can be a blessing for future generations. This commitment will be a source of happiness and fulfilment.

◄ *Permaculture garden on an island in the Bahamas.*

▼ *Fragile paradise: the sublime environment of eastern low coral islands is threatened more by the rise of tourism than by the rise of oceans and the death of corals. Will we wait without reacting to the destruction of the biosphere?*

◄ *The ethics of permaculture invites us to take care of the Earth, but also human beings. Living in a permaculture garden, we try to provide our land with the best possible care, and it does well for us in return.*

28

TAKE CARE OF THE FARMER

Choose your words carefully, because they are what creates the world around you.

Navajo wisdom

This is the penultimate chapter of this manual, which I approach with humility because it is, like so many other aspects of this book, nourished by our errors more than our successes. We have both neglected our own needs during this adventure. We would have spared ourselves a lot of suffering if we had been more attentive to caring for ourselves and listening to others.

Because creating a farm, as you will see, is an exhilarating, but difficult, adventure. It can push you to your limits. But should these difficulties force us to give up? Certainly not! We are stronger for overcoming them. However, to be able to live the adventure joyfully, it is better to lighten the load by avoiding predictable traps.

Making life easier requires some very concrete things like good posture, more efficient tools, good organisation... All of the technical tips offered in this manual have no other purpose than to allow you to succeed in your project more serenely by freeing up time and avoiding fatigue and worries.

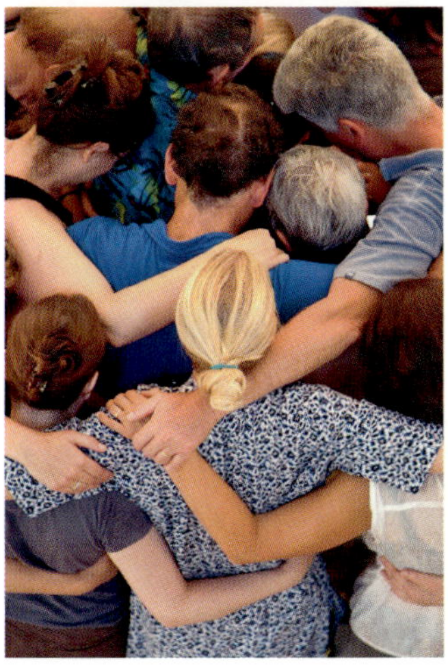

Here are some tips from our experience, very naive certainly, but sincere. I have battled and often failed on just about every point mentioned below... I wish you good luck from the bottom of my heart to do better. Write your own story and make it as happy as possible!

Tune into your Deeper Self

Before launching into creating a farm, make sure that this project is truly yours, the one that suits you. That you are in your rightful place, in your true 'ecological niche'. Things will flow much smoother this way.

When we deeply love what we do, we often have the feeling of not working, but rather having fun!

Respect Those who are Committed to Your Side

It is very likely that your adventure will not be solo. Other people will participate in one way or another. Is this a couples project? Will you take your children on board for the adventure?

Be deeply attentive to the dreams of others, especially that of your life partner. Their dreams are different to yours. Respect their uniqueness. Give them some space to express themselves. An Indian proverb says: "Whatever you don't do along with others, you do against them."

Take Care of your Family's Equilibrium

Cultivate your romantic relationship. Nothing is ever a given; love, like plants, needs water and nourishment. Go on a date as a couple often. Have common interests outside of work. Make love divinely.

Go to the swimming pool or to the cinema with your children. Play and laugh with them. In the evening, your place is not in the greenhouse with a headlamp, but alongside the young ones, to help them with school work or tell them stories. Be completely available for the older ones who are searching for their way in life.

▲ *Laure Lemarchand, lawyer and sustainable development consultant, has become a wonderful yoga teacher. She leads annual ecology workshops at the farm.*

Take your Time

We cannot emphasise enough the importance of starting small and allowing your project to grow organically. We do not grow a plant by pulling on it. Do not give birth to your project with forceps. Give it time to hatch. Small is beautiful!

Design Carefully

Spend the time necessary on your initial design, then constantly improve it. Be on the lookout for anything that will make your life easier, allow you to save time, by limiting effort, unnecessary trips and more. Fifteen minutes saved every day thanks to a few small improvements equals two weeks of vacation at the end of the year.

Preserve your Health

Your body will be stressed like a professional athlete. The life of a market gardener is often a marathon where you have to go the distance. We all know the rules of a healthy lifestyle, yet how easy it is to neglect them! Like any living organism, our body needs good nutrition and balanced rhythms. You cannot hold on forever in over exertion; you need to allow yourself ample recovery time. Have regular meals and sleep schedules. Give yourself a break, ideally at least one day per week, even if the workload is great. You will be more effective this way. Take a few days of vacation whenever possible.

Practise Garden Yoga

Listen to your body while gardening. By forgetting to pay attention to our bodies, some market farmers end up visiting an osteopath. Practising yoga, qi gong or tai chi contributes powerfully to nourishing the awareness of our movements. Our efforts must be judicious and not put us in pain.

This requires adapted postures. To lift heavy loads, bend your legs and keep your back straight. Avoid working bent over for too long, or posture becomes bent forward. Search for the most comfortable posture for each type of task, the least tiring for you. Inhabit your body.

LOVE, PEACE

MAY THE EARTH BE WITH YOU

UNDERSTANDING TOGETHER

Choose the Most Suitable Tools

Get yourself some great tools and know how to use them. Use the right tool for each operation, the one that will allow you to work quickly and stand with your back straight when possible. This is generally the largest tool available according to the task at hand, with a long handle rather than a short one.

Choose lightweight tools – when a handle weighs a few hundred grams less, it saves a lot of effort when a task is repeated thousands of times.

Comfortably arrange your workspaces. There is no shame in planting seeds sitting down instead of always standing up!

Breathe

It's a good idea to take a break every hour, just to straighten up. Take a deep breath and become aware of your body, look at the surroundings, listen to the birds sing, and come back to the present moment. This minute of returning within oneself can become a little ritual.

Cultivate your Interior Garden

When you are alone all day in your garden, hands work and ideas come and go like the clouds in the sky... Happy clouds, clouds loaded with anger or anguish sometimes... Welcome them but don't let them settle. Weed out toxic thoughts before they invade your mental space, and cultivate positive thoughts. It is a form of meditation. If you are inclined towards contemplation or spirituality, these hours of solitude facing inward in the great outdoors are treasures.

Curiosity and Enthusiasm

Don't become a hand-planting machine. Have interests unrelated to your work.

Stay Connected

Don't isolate yourself. Cultivate the art of living with others. Give and receive – the more you give, the more you receive. If that doesn't work, have the wherewithal to ask for help.

Be Brave

When life gets really difficult, don't think of yourself as a victim; you were warned and you are fully responsible for your actions. Taking responsibility for a situation gives us power over it. Be brave; keep your head up. Nowadays courage is not taught, yet each of us has it. Adversity will lead you to discover unsuspected resources within you.

▲ *Élodie, Vanessa and Cécile, facilitators of our courses and market gardeners in the making, pictured in an acroyoga session at the pond. This type of fun practice develops confidence in oneself and in others.*

Be Flexible

Your project will be constantly evolving; perhaps it will lead you to paths that you did not expect? Do not allow yourself to be locked into a project that no longer suits you. React; take back the helm. Revisit your design!

Express your Gratitude

To be a farmer is to be a guardian of a small piece of the only known living planet. What an honour! Don't let the little worries of everyday life overwhelm you; reach for the sky. Don't lose sight of the meaning of your commitment and be grateful for what you have.

Love What Is

Ask Mother Earth to teach you wisdom. We are alive on this Earth today and gone tomorrow... All or almost all leaves us, so let's give the best of ourselves and relax. Trust!

▲ *An ephemeral mandala created by our trainees, an invitation to connect with others and nature... But is it necessary to make this distinction? Are we not part of nature?*

Be happy on the days when the weather is nice, and be happy on frosty or stormy days. Ideal conditions do not exist. Love what is!

Inner Ecology

My commitment to farming finds its roots in the happy memories of my childhood, but also among the native people who taught me so much. Like them I think that all of life is one. It unfolds in a festival of forms and colours, but it is the same principle with carrots and ladybirds of the garden, my horse or my wife. Taking care of all these forms of life is a noble mission. Approaching old age now, I am filled with infinite respect for this dazzling life, amazed at the miracle of being here, inhabiting Planet Earth.

► *Three boards make a raft for the doll and Fénoua is off on a journey... Let's keep our inner child's heart intact and our ability to dream with eyes open!*

Is there a true ecology that does not start from deep within oneself? Our inner world is a reflection of the cosmos. If we cultivate values of empathy and respect, they transform the world that surrounds us.

We have a workshop on the farm that is particularly close to my heart. It is called 'Inner Ecology'. Using yoga and therapeutic massage as vectors, we learn to connect with our inner being, each other and nature, in a climate of unconditional acceptance and empathy. Each participant comes out transformed. Human beings only want to be loved as they are, live in accordance with their values and with the cosmos. It's simple: if we are at peace with ourselves, we will stop destroying nature.

Talking to Plants and Animals

In the Amazon, my shaman friends speak with plants. This fascinates me. I am convinced that we can reach a level of communication with plants and animals that is infinitely deeper than the superficial relationship that we are accustomed to in our materialistic society. Try talking with your plant and animal companions, and you will see it for yourself! They speak to you; can you hear them? These dialogues are difficult to describe because they do not pass through ordinary channels. They live with the heart and soul, beyond words. The messages are imprinted on us like intuitions, or forms of energy that are clear and immediate.

There is the world of matter that we see, the garden made of shapes and colours in which we work, and then there is an invisible, vast world that we can perceive with our inner eye. Your visible garden is accompanied by an invisible garden made up of subtle energies that interact with ours. By listening, you will be able to feel these energy fields.

All these visible and invisible beings that surround us are sensitive to the quality of our intentions and energy. Our intentions are a powerful message. Garden with an open heart and offer the Earth the best of yourself. You are not alone! Be sure they are working on your side. We have the same goal to promote the flourishing of life.

It is much easier to feel the energies of the garden when working by hand in silence, rather than with a tractor and sophisticated machines.

Go Forward with an Open Heart and Confidence

The revolutionaries of the 20th century fought for a better world with Kalashnikovs and Stalin's organs (Katyusha rocket launcher). We seek to make the world we dream of possible by planting apple trees and growing radishes and tomatoes.

Human beings are part of nature; however, we have a very particular responsibility towards animals and plants. We are a link between Earth and Sky, between the visible and the invisible. The special gifts that we have received allow us to take care of our companions.

If the world is doing so badly today, it's because human beings have lost sight of this duty. They behave as if the Earth were a product to consume. They forget that she is their mother and destroy the biosphere without understanding that, in doing so, they destroy themselves. These attacks on nature also tear apart the bonds of brotherhood between us, plunging entire sections of humanity into precariousness and despair.

But things are beginning to change quickly. We are witnessing the birth of a new type of humanity, conscious of the fact that the only truly precious product is life, in all its forms. There is not a single village on this planet where women, men, and children do not stand up and engage in the preservation of life.

The biosphere is deteriorating so quickly that nothing is guaranteed. To tell the truth, no one knows if we are heading towards a collective suicide or a profound regeneration of the planet and humanity. This fate is in the hands of each of us. Each human being, even the most humble person, holds incredible power. No need to be strong, educated, in good health or fortunate to take up the cause, because each of us, deep down, is connected to an infinitely powerful energy that is available for one thing – that we make ourselves available, that we commit our lives to repair, to heal, laugh, embellish, and connect. This can be done in a thousand and one ways, depending on the unique qualities of each one of us.

Creating a garden, even a tiny one, is a powerful commitment. Each organic garden, even the smallest, improves soil health to protect biodiversity. It feeds a few people with healthy fruits and vegetables, develops bonds of solidarity within the local community while helping to heal the atmosphere. It's quite a nice challenge.

There is no need to ask if we will win this battle in order to engage in it. Be sure of one thing: our sisters and our brothers with leaves, needles, fur, feathers, and scales are waiting for us to become aware of our responsibility towards this living planet. Be confident that we are making a commitment to a living world that has overcome many threats for nearly four billion years. We are not alone in this peaceful and noble endeavour: a powerful current from the depths of ages carries us! I have the deep conviction that by fully accord with nature, we can repair the biosphere much faster than we might think.

The world of tomorrow is in our hands. Let's move forward joyfully!

29

THE MOST BEAUTIFUL GARDEN IN THE WORLD

Free-spirit, free universe.

Koa[103]

It is up to you to write the final chapter of this manual. Our deepest desire is that these pages help you to realise the adventure of your life.

Your garden will be the most beautiful because it comes from your dreams and the work of your hands.

Best wishes and happiness from the bottom of our hearts!

◂ *The mandala garden on an April morning.*

CONCLUSION

Dear friends,

The production of this manual took place over six long years. The closer we get to the end, the more aware I become of its imperfections. I request your indulgence in the faults of this book. The world of ecologists is too often a struggle against one another, yet it is by uniting our strengths and talents that we will save the biosphere. I appeal to your kindness: let's move forward together to explore these new fields of knowledge; let's share our discoveries!

Ecoculture is like a relay race: we are motivated by its many pioneers, and this manual is our contribution; make yours by carrying the baton further.

We are grateful to readers who share their comments and experiences with us.

We would like to add one last thing before leaving you: Perrine and I have two options for our future. Either we respond to emails and calls that we receive by the thousands, give conferences, and participate in media events, or we return to our garden beds to practise what we presented here in this manual and continue to make progress. We have chosen to remain small farmers. Bec Hellouin Farm is a small family farm with a small team and we are not equipped to respond to the endless stream of requests and solicitations. Forgive us if your emails and letters do not always receive an answer, but know that we read them, but sometimes we keep the farm gate closed. It is only open during the times indicated on our website.[104]

We wish to dedicate ourselves resolutely to the implementation of our own ecological transition. Since we started working on this manual, we have seen the world transform profoundly. Climate change is accelerating and becoming more evident each year - we are noticing it through the effects on the farm. However, humanity is not taking the measures that this ecological emergency warrants.

The collapse of our civilisation no longer seems questionable, but rather a certainty. A civilisation that consumes so much more than the planet is capable of, cannot endure. When will this collapse occur, and what form will it take? It is impossible to answer these questions. But there is one certainty: the more we delay in acting, the more brutal the shock will be.

Faced with these observations, we ask ourselves: what are we doing to reduce our ecological footprint? In recent weeks, we have had many conversations with our children. They made a chart called 'Our family commitment to the planet' where we note our intended actions. We decided together to move things up a step by reducing our ecological footprint to a single planet - isn't this the one unique solution? Ideally, we want our home to disconnect from water and electricity networks and become self-sufficient. And starting next year we will try to eliminate all fossil fuel usage within the farm.

As rural folk, we are committed to self-sufficiency. This evolution is certainly easier to carry out here than in the city where garden space is limited. In the event of a sudden collapse how will city populations be fed? To achieve true food security, we need to connect all of our towns and counties with millions of microfarms. Farms that operate efficiently without oil. Obviously, this is not going to happen overnight but given the urgency of this crisis, we will have to begin today to prepare for a gentle transition from one way of life to another.

The more we ponder and observe transition initiatives around the world, the more we believe that the best alternative to technologies is an alternative to technologies. Artificial intelligence turns out to be terribly energy-hungry. Let's not turn our backs on technological and scientific progress, but use it with awareness and discernment. In the event of a major collapse, most of our technologies will become inoperative.

Learn the simple ways to use renewable biological resources and manage them sustainably. It is our entire way of life that needs to change. This is in no way a punishment, but rather a beautiful adventure that requires creativity, courage, total commitment and dedication to using all of our talents. It will make us grow as human beings. We need to create autonomous, fertile and resilient living spaces that will become lifelines if everything changes all of a sudden. A united community around each garden, each farm, each neighbourhood, connected by the invisible links of sharing, with emphasis on the greater good.

We passionately love our era and wouldn't want to go back in time. Respect for others and their uniqueness, a taste for freedom and democracy, recognition of women and the values that they hold, knowledge of the biosphere and awareness of its beauty, the desire to share fairly the resources of our common home grow every day, particularly among young people.

The ecological crisis is a factor in accelerating our evolution. Could it be that human beings are in the process of emerging as a species?

More and more of us are becoming aware that a material vision of our existence cuts off an essential part of ourselves. Take care of Life in all its forms, feel connected to nature, to one's human sisters and brothers, to one's inner self, to give the best of oneself for a cause that is greater than ourselves: is this not the essence of all forms of spirituality?

The looming collapse offers us a unique opportunity to build a better world, unprecedented in the history of humanity. Let's influence the situation rather than suffer from it; let's make it desirable and joyful. Without waiting until it is too late, let us reconnect with beautiful nature in all of its infinite variety and generosity. Let's adopt a simple and sustainable lifestyle; certainly, we will no longer have gadgets, but we will have an abundance of what makes life beautiful and serene!

Our best wishes for your success!

Bec Hellouin Farm, November 19, 2018

ENDNOTES

1. Blanche Magarinos-Rey, *Semences hors la loi*, 'Manifestô', Alternatives, Paris, 2015, p.9.
2. This chapter is based on a synthesis of these documents:
   Christian Boué, *Produire ses Graines Bio, Légumes, Fleurs et Aromatiques*, Terre Vivante, Mens, 2012.
   FNAB, Bio d'Aquitaine and CAB, *En Bio, Semons la Diversité! Recueil d'expériences du Réseau FNAB sur les Semences de Population en Grandes Cultures*, 2014.
   Jérôme Goust, *Le Plaisir de Faire ses Graines. Un Guide pour Retrouver son Autonomie au Jardin*, Éditions de Terran, Escalquens, 2005.
   Andrea Heistinger, Arche Noah and Pro Specie Rara, *Semences Potagères. Le Manuel pour les Produire soi-même*, trad. Sylvie Girard-Lagorce, Rouergue, Arles, 2015.
3. FNAB, Bio d'Aquitaine and CAB, *op. cit.*, p.7.
4. Andrea Heistinger, Arche Noah and Pro Specie Rara, *op. cit.*, p.16.
5. *Ibid.*, p.29.
6. Blanche Magarinos-Rey, *op. cit.*, p.8.
7. www.andrea-heistinger.at/manual-seed-saving
8. www.upov.int/portal/index.html.fr.
9. Andrea Heistinger, Arche Noah and Pro Specie Rara, *op. cit.*, p.34.
10. The procedure is particularly well detailed in the books by Christian Boué (*op. cit.*) and Andrea Heistinger (*op. cit.*). The paragraphs in this section are based on the testimonies of these authors, as well as the experiments carried out at Bec Hellouin Farm.
11. Maurice Chaudière, *La Forêt Fruitière, Éditions de Terran*, Sengouagnet, 2008, p.12.
12. Benoît Priel, Denis Retournard and Robert Elger, *Le Petit Traité Rustica pour Semer, Greffer, Bouturer, Rustica*, Paris, 2015.
   Then, to go further: Maurice Chaudière, *De Greffes en Greffes. La forêt fruitière, l'art de rendre productifs friches, landes, causses, garrigues et maquis*, Éditions de Terran, Escalquens, 2016.
13. We recommend: Alain Pontoppidan, *J'apprends à Greffer mes Arbres Fruitiers et mes Arbres d'ornement*, Terre Vivante, Mens, 2013.
14. John Seymour, *op. cit.*, p.50.
15. *Ecophyto. Portrait de Jean-Philippe Deguigne, Agro-écologue à la Réunion*, YouTube, www.youtube.com/watch?v=nA4sokvIGiM.
16. Janine M. Benyus, *Biomimétisme. Quand la nature inspire des innovations durables,* Rue de l'Échiquier, Paris, 2011, p.38.
17. www.franceinter.fr/emissions/interception/interception-13-mai-2018.
18. See on this subject the work of Éric Petiot, 'Les Plantes pour Soigner les Plantes', *Viticultures Alternatives*, 29-30 October 2015, www.youtube.com/watch?v=9YcFw53JEzc.
19. Francis Chaboussou, *Les Plantes Malades des Pesticides*, Debard, Paris, 1980, cited by Éric Petiot, *op.cit*.
20. *Ibid.*
21. *Ibid.*
22. Éric Petiot, Bernard Bertrand and Jean-Paul Collaert, *Purin d'ortie et Compagnie, Éditions de Terran*, Escalquens, 2012 (4th edition.).
   See also www.youtube.com/watch?v=RcgmNx5HudU.
   The website www.purindortie-bretagne.com offers quality products and recipes to reproduce at home.
23. Gildas Véret, *Permaculture, Créer un Mode de vie Durable*, Rustica, Paris, 2017, p.29.
24. www.goodplanet.info/photo/2017/10/13/sixieme-extinction.
25. Johanna Villenave-Chasset, *Biodiversité Fonctionnelle. Protection des Cultures et Auxiliaires Sauvages*, France Agricole, Paris, 2017, p.16.
26. Blaise Leclerc and Gilles Leblais, *Des Auxiliaires dans mon Jardin! Les Attirer, les Loger, les Nourrir*, Terre Vivante, Mens, 2015, p.92.
27. Camille Joyeux wrote this note based on the following sources:
   CPIE de la Bresse et du Jura, *Quelques aménagements simples pour transformer un espace en refuge naturel*, jardiner-naturellement.org.
   Andréa Haug, '421 Millions d'oiseaux Européens ont Disparu en Trente Ans', *Futura Planète*, 2014, www.futura-sciences.com/planete/actualites/ oiseaux-421-millions-oiseaux-europeens-ont-disparu- 30-ans-55955.
   Gilles Leblais, *J'accueille et j'observe les Oiseaux dans mon Jardin*, Terre Vivante, Mens, 2016. www.youtube.com/watch?v=clohvt-zqwg.
   LPO, *Aménagez votre Jardin pour Accueillir les Oiseaux*, www.lpo-boutique.com.
   LPO, catalogue 2017. LPO, *Liste Rouge des Oiseaux d'Europe: un constat alarmant*, 2015, www.lpo.fr/actualites/liste-rouge-des-oiseaux-d-europe-un-constat-alarmant.
   LPO PACA, *Aider les oiseaux à nicher*, paca.lpo.fr.
   nichoirs.net/index.html.
   www.ornithomedia.com.
   Facebook Page of the Miracle Farms.
   Vivara, www.vivara.fr.
28. Gustave Flaubert, *Madame Bovary*, 1857, cité par Philippe Madeline and Jean-Marc Moriceau, *Les Paysans*, Les Arènes, Paris, 2012, p.10.
29. According to the FAO, "the livestock sector contributes to 14.5% of greenhouse gas emissions from human activity'. FAO, 'L'élevage et l'environnement", www.fao.org/livestock-environment/fr.
30. Harvey Arden and Steve Wall, *Les Gardiens de la Sagesse,* Éditions du Rocher, Monaco, 1994, p.63.
31. Marcel Mazoyer and Laurence Roudart, 'Des Agricultures Manuelles à la Motorisation Lourde: des Écarts de Productivité Considérables', *Grain de Sel*, n°48, 2009, p.12.
32. Jean Nolle, *Machines Modernes à Traction Animale. Itinéraire d'un Inventeur au Service des Petits Paysans*, AFDI-L'Harmattan-gret, Paris, 1989, p.463.
33. PROMMATA, La Gare, 09420 Rimont, www.assoprommata.org.
34. www.prommata.org and www.assoprommata.org.
35. H2C, 63290 Châteldon, H2c-distribution. over-blog. com.
36. Harvey Arden and Steve Wall, *op. cit.*, p.51.
37. Comptoir de la Faux, www.comptoirdelafaux.com/topic/index.html.
38. L'École du Pain des Maîtres de mon Moulin, www.ecoledupain.com.
39. Quoted by Pierre Sauvageot and Paul Grillo, *La Culture Familiale du Blé, Facteur de Renaissance Française*, Sequana, Paris, 1943.
40. L'École du Pain des Maîtres de mon Moulin, www.ecoledupain.com.
41. *Maison Rustique du XIXe Siècle. Encyclopédie d'agriculture Pratique*, La Maison Rustique, Paris, 1835, pp.330-331.
42. *Ibid.*, p.331.
43. To learn more: Deart, S'installer Paysan Boulanger, Fiche Technique, 2012, www.agriculturepaysanne.org/files/s-intaller-paysan-boulanger2.
   Guillaume Le Du, 'Tanguy Fait du Pain avec la Farine de Son Blé', *Ouest-France*, numéro spécial *Avenir-formation-metiers*, 30 Mars 2017.
   CFPPA de Montmorot: www.montmorot.educagri.fr/cfppa/paysan-boulanger-fournier.
   CFPPA de Coutances: www.coutances.educagri.fr/cfppa_form_bprea_pb.php.
44. Joseph Pousset, *Traité d'agroécologie, pour une Agriculture Naturelle*, France Agricole, Paris, 2012.
45. Maison Rustique du XIXe Siècle, *op. cit.*
46. Pierre Sauvageot and Paul Grillo, *op. cit.*, pp.24-27.
47. For more information: 'Faire un Max de Blé Grâce à la Permaculture: la Méthode Fukuoka-Bonfils', *Jardinons la Planète*, 2013, jardinons.wordpress.com/2013/01/19/faire-un-max-de-ble-grace-a-la-permaculture-la-methode-fukuoka-bonfils.
48. Ivan Illich, *La Convivialité*, Seuil, Paris, 1973, p.43.
49. *Ibid.*
50. La Fabriculture, www.lafabriculture.fr.
51. Bernard Bertrand, *Guide du Travail Manuel du Bois à la Plane et au Banc à Planer*, Éditions de Terran, Escalquens, 2016.
52. Azby Brown, *The Genius of Japanese Carpentry*, Tuttle Publishing, Clarendon, 2013, p.74.
53. Philippe Desbrosses, *Nous Redeviendrons Paysans*, Dangles, Escalquens, 2014, p.21.
54. Imane Khireddine-Medouni, Éléonore Breuillard, Claire Bossard, 'Surveillance de la Mortalité par Suicide des Agriculteurs Exploitants Situation 2010-2011 et évolution 2007-2011', Santé publique France, 2016, p.5.
55. Gérard Cornu, 'Taux de suicide chez les agriculteurs', January 19th 2017, question écrite no 24706, p.137.
56. www.fnab.org/loi-dorientation-agricole-il-faut-enseigner-la-bio-pour-susciter-des-vocations
57. The useful (or used) agricultural surface (or surface area) is a standardised concept in European agriculture. It is intended to evaluate the territory devoted to agricultural production. It is considered arable land (field crops, market gardens, temporary meadows, fallows, crops under shelter and allotment gardens, etc.), of surfaces always kept as grass (permanent meadows, mountain pastures), perennial crops (vineyards, orchards, etc.). It does not include woods and forests. According to Actu-environnemment.com, 'Useful agricultural area'.
58. Anna Lee Walters, *The Spirit of Native America*, translation Danièle and Pierre Bondil, Casterman, Tournai, 1990.
59. The FNAB (National Federation of Agriculture organic) consists of a national federation and regional and departmental groups distributed in the territory. www.fnab.org.
60. The Chambers of Agriculture intervene in the field with farmers, agricultural employees, foresters and communities on all questions of agricultural interest. www.chambres-agriculture.fr/chambres-dagriculture/know-us/the-network-of-chambers-of-agriculture
61. "The ADEAR are the Associations for

Development of Agricultural and Rural Employment [...]. An ADEAR is an association that brings together peasants, the majority members of the Peasant Confederation, and other actors from around the world by the desire to share their experience and their know-how to help maintain and to settle numerous peasants and to support the values of peasant agriculture. www.jeminstallpaysan.org/qui-sommes-nous.

62. The CIVAM network (Centres of Initiatives to Promote Agriculture and the Rural Environment) is an associative actor of agricultural and rural development, which has been operating for more than fifty years.
63. "The Inpact network is an associative platform resulting from the merger of agricultural associative networks": www.agricultures-alternatives.org/rubrique12.html
64. Biofarm: An association that helps the development of small agroecological farms, www.biofermes.fr.
65. Some countries have regulations on diverting water courses. Make sure you've researched your area.
66. Nicolas Soulier, *Reconquérir les Rues*, Ulmer, Paris, 2012, p.7.
67. Perrine and Charles Hervé-Gruyer, *Miraculous Abundance*, Chelsea Green Publishing, White River Junction, 2016.
68. Roger S. Ulrich, 'View Through a Window May Influence Recovery from Surgery', *Science*, 224, 1984, pp.420-421.
    Seong-Hyun Park and Richard H. Mattson, 'Therapeutic Influences of Plants in Hospital Rooms on Surgical Recovery', *HortScience*, 44, 2009, pp.102-105.
69. Fritjof Capra, *Le Temps du Changement*, trad. Paul Couturiau, Éditions du Rocher, Monaco, 1983, p.382.
70. Dhyani Ywahoo, *Sagesse Amérindienne*, trans. Michel Saint-Germain, L'Homme Éditions, Montréal, 1999.
71. Olivier De Schutter, *L'Économie Politique de la Faim. Garantir le droit à l'alimentation dans un monde de ressources rares*, Groupe ESA, Les leçons Inaugurales du Groupe ESA, Angers, 2010, p.7, docplayer.fr/868963-L-economie-politique-de-la-faim.html.
72. www.wwoof.fr/accueil, www.msa.fr/lfy/embauche/wwoofing, www.terre-net.fr/observatoire-technique-culturale/reglementation-social-juridique-fiscal/article/le-wwoofing-du-benevolat-sans-statut-220-124075.html
73. Camille Joyeux and Louise Géhin, *Sol et Fertilité*. Analyses des sols de la Ferme biologique du Bec Hellouin par Claude et Lydia Bourguignon, Institut Sylva, 2017, www.fermedubec.com/la-recherche/les-rapports-scientifiques., p.11.
74. Pablo Servigne, *Nourrir l'Europe en Temps de Crise*, Nature et Progrès, Jambes, 2014, p.101. Rééd. Actes Sud, Babel no 1499, Arles, 2017.
75. On the DGCCRF website, for example, simply enter the word 'jam' in the search bar and you come across this page with practical sheets: www.economie.gouv.fr/dgccrf/Publications/Vie-pratique/Fiches-pratiques/Confitures-gelees-marmelades-de-fruits
76. See www.alimentaire-pro.com/dossiers/conservation/sterilisation/laboratoires.php.
    On the website of the Ministry of Agriculture and Food, there are guides to good hygiene practices, some of which are accessible online: agriculture.gouv.fr/liste-des-guides-de-bonnes-pratiques-dhygiene-accessibles-en-ligne
    Complete list of guides: agriculture.gouv.fr/sites/minagri/files/liste_gbph_francais_publies_avril2016.pdf.
    Concerning the regulations regarding health matters during processing on the farm: agriculture.gouv.fr.
    Agrobio Poitou-Charentes offers a guide for producers wishing to offer a workshop: www.penser-bio.fr/img/pdf/Fiches-filieres_transfo_a_la_ferme-2.pdf.
77. There is a lot of information on processing at www.chambre-agriculture-61.fr/gestion-de-lexploitation/agritourisme-et-circuits-courts/transformer-ses-produits/transformer-des-fruits-et-legumes.
    This article provides food for thought for design of a product transformation workshop: www.alimentaire-pro.com/dossiersconservation/sterilisation/laboratoires.php.
78. The composition of your jam impacts its name: it cannot be called 'jam' if the weight of sugar used is greater than 55%. A jam 'extra' must be prepared with more than 450g of fruit pulp per kilo of final jam. Terms 'jelly', 'marmalade', 'cream' or 'candied' are also regulated (decree of August 14, 1985 as amended). www.agriculteur-normand.com.
79. The maximum levels are as follows:
    - 50 micrograms per kilo for fruit juices (including those reconstituted from concentrates), fruit nectars, spirits and ciders and others drinks produced from apple juice and apples;
    - 25 micrograms per kilo for solid products made from apples (compotes, purees) intended for direct consumption;
    - 10 micrograms per kilo for juices and products with a base of apple pieces;
    - 10 micrograms per kilo for food (other than those prepared from cereals) intended for infants and toddlers, and labelled and sold as such.
    Regulation (EC) No 178/2010(9) for the levy and comprehensive food sample analysis; the previous Regulation (EC) No 401/2006(10)
    www.anses.fr/fr/system/files/mic2011sa0037Fi.pdf.
80. The minimum acetic content of vinegars is 5 grams of acetic acid per 100 millilitres.
    However, this minimum is 6 grams of acetic acid per 100 millilitres for wine vinegars. The acetic content of vinegars, expressed in acetimetric degrees, is equal to their total acidity expressed in grams of acetic acid for 100 millilitres of vinegar measured at a temperature of +20°C. A difference of 0.2 degrees, or 2 grams of acetic acid per litre of vinegar, can be admitted, to the extent of this content: www.legifrance.gouv.fr/affichTexte.do?cidTexte=jorftext000000867411&dateTexte=20170315.
    In addition: "For the marketing of products under the name vinegar, two conditions must be respected regulatorily:
    1. The minimum degree (%) of acetic acid (indicated on labelling)
    - Minimum 6° for wine vinegars.
    - Minimum 5° for other vinegars (mead, cider, fermented juices, fruits other than grapes).
    2. Residual alcohol content
    - The residual alcohol content of wine vinegars is limited to 1.5% by volume.
    - The residual alcohol content of vinegars other than wine vinegar is limited to 0.5% by volume.
    - For vinegars from liqueur wines or natural sweet wines, this content can reach 3% in volume": www.cevennes-ecotourisme.com/img/pdf/support_vinegars.pdf.
81. Eliot Coleman, *Des Légumes en Hiver*, trans. Elsa Petit, Actes Sud, 2019 (2021), p.157.
82. Excerpt from Market Gardening. Agricultural Production Guideorganic in Aquitaine. Marketing, 2013, www.nouvelle-aquitaine.chambres-agriculture.fr.
83. See the guide, d'Ecocert, for the use of logos: www.ecocert.fr/files/ID-SC-172-guide-pratique-etiquetage-12.12.12.pdf.
    For more specific information regarding products subject to specific standards: eur-lex.europa.eu/legal-content/fr/txt/?uri=celex%3A32008R1221.
84. https://rnm.franceagrimer.fr/prix?M0201:marche
85. Eliot Coleman, *The Winter Harvest Handbook*, Chelsea Green Publishing, 2009, p.157.
86. You can also contact the Associations de Formation Collective à la Gestion (AFOCG), if you want to do your own accounting; for the price of a membership to the association.
87. Source: www.youtube.com/watch?v=-_luwqeiA6Y.
88. Marie-Claire Ricard and Hervé Monestier, *La Ferme de Notre Enfance*, De Borée, Sayat, 2012, p.13.
89. Lanza del Vasto, *Technique de la Non-violence*, 'Folio essais', Gallimard, Paris, 1988, p.71.
90. Perrine and Charles Hervé-Gruyer, *op. cit.*, p.266.
91. Dhyani Ywahoo, *op. cit.*
92. Pablo Servigne, *Nourrir l'Europe en Temps de Crise. Vers des systèmes alimentaires résilients*, Actes Sud, Arles, 2017.
93. Pablo Servigne, personal communication.
94. Eliot Coleman, *Des légumes en hiver*, *op. cit.*, p.157
95. Camille Joyeux, Charles Hervé-Gruyer and Louise Géhin, *Peut-on Vivre d'une Forêt-jardin?*, Institut Sylva, rapport technico-économique no 1, 2017, www.fermedubec.com/la-recherche/les-rapports-scientifiques
96. These notes are an excerpt from report n.5 from the research programme, *Maraîchage biologique permaculturel et performance économique*, available on our website: Sacha Guégan, François Léger, Charles Hervé-Gruyer and Gauthier Chapelle, *op. cit.* www.fermedubec.com/la-ferme/la-recherche/#rapports-scientifiques
97. Louise Géhin, François Warlop and Cécile Thibaut, *op. cit.*
98. Lanza del Vasto, *Technique de la Non-violence*, 'Folio essais', Gallimard, Paris, 1988, p.64
99. Ivan Illich, *La Convivialité*, *op. cit.*, p.128.
100. Marcel Mazoyer and Laurence Roudart, *op. cit.*
101. Laurence Estival, 'Pour Nourrir le Monde: l'Agriculture familiale', *Alternatives Économiques*, no 338 bis, September 2014, pp.4-6.
102. See the work done by La Maison de l'Artemisia, created by our friends Lucile and Henri de Cossé-Brissac, who demonstrated the effectiveness of this plant as malaria treatment without side effects: maison-artemisia.org.
    Also see Lucile Cornet-Vernet, *Artemisia. Une plante accessible à tous pour éradiquer le paludisme*, 'Domaine du possible', Actes Sud, Arles, 2018.
103. Paroles zen, text provided by Marc de Smedt, Albin Michel, Paris, 1996, p.37.
104. www.fermedubec.com contains various information and news programme reports for scientific research taking place on the farm are presented there.

APPENDIX TABLES

TOOLS USED AT BEC HELLOUIN

In this appendix we present the tools used in the intensive heart of our farm (they don't include animal traction tools – those are described in Chapter 7, p.71).

These tables also include the equipment needed for sales as we do it at Bec Hellouin Farm (sales to individuals in the form of baskets, and restaurateurs). Some tools are used for different activities (harvesting, storage, transport, marketing, etc.). They appear a few times in the different tables presented.

Information about suppliers and prices given here will save a lot of time for future market gardeners. These lists were produced 10 years after the creation of the farm; it is possible to start with much less equipment, of course.

The prices indicated in the tables are for information only. These are before tax prices, found on the internet, in professional catalogues or specific suppliers for the year 2017. They do not include shipping and installation costs.[3] Future maintenance costs and consumable stock replenishment are not included in the estimates.

Soil Preparation

1. Decompacts and aerates soil without turning it;
2. Refines soil and prepares the bed for sowing or transplanting.

The tools used and the procedure to follow are different depending on bed type:

- For flat beds: soil preparation includes decompacting, levelling and refining to create optimal conditions for use of the Eliot Coleman precision seeder. For this procedure, several tools are used: campagnole, microcultivator if necessary, an 80cm wide rake, a classic rake.
- For rounded beds: decompact the soil, reform the rounded bed and refine the soil. A broadfork, hoe and rake are in turn used for this purpose.

TOOLS	PRICE OR AVERAGE PRICE	NOTES
Campagnole[4]	Campagnole allows the soil to be decompacted in depth without turning it over and disturbing the surface horizons and soil life. It also helps refine the seedbed: - Rigid counter spaced tines spaced 10cm apart (compact and clay soils): €285 incl. tax - Flexible counter-tines spaced 8cm apart (aerated and/or stony soil – finer crumbling): €330 including tax	Prepares flat beds
Microcultivator[5]	$495 (€427 approx.) Coleman tool: the microcultivator offers superficial work of the first 5cm of soil. It allows you to crush the roots of the previous crop, but also to decompact the soil without turning and without surfacing weed seeds. The tool works with a drill. Buy a cordless drill (and spare battery) sufficiently powerful – at least 18V (pay attention to the amperage).	Prepares flat beds
Broadfork[6]	With four teeth: €95 incl. tax, i.e. €75.84 excl. tax	Rounded beds
Rakes[7]	- Classic rake 14 teeth, straight or curved teeth: €17 - 17-tooth aluminium rake with tilting head: €76 - Lawn rake with 44 inclined teeth: €41 - 80cm rake: for preparing the seedbed, it allows you to remove stones and debris but also to level the soil for a smooth and perfectly homogeneous seedbed: $85 (€73 approx.).	Prepares flat and rounded beds
Hoe[8]	Bronze hoe from PKS (950g): €72 incl. VAT	Rounded beds
String	String with winder (100m string): €13	Preparation of flat beds, use two per bed Design a system for storing, rolling up and unwinding the string quickly and without tangling Self-made: string and two concrete irons

1. In order to reduce shipping costs, it is advisable to order American tools without their handles.
2. More information on the Campagnole visit: www.lafabriculture.fr, www.lafabriculture.fr/la-campagnole-89.html, www.lafabriculture.fr/img/pdf/campagnole_maraicher_.pdf.
3. More information on the microcultivator: www.johnnyseeds.com/p-6620-tilther.aspx.
4. More information on the broadfork: https://graines-grelin-grelinette.fr.
5. More information on the 80cm rake on the Johnny's Seeds website: www.johnnyseeds.com/p-5919-bed-preparation-rake.aspx.
6. More information on the bronze hoe from PKS: www.kupferspuren.at/fr.html, www. kupferspuren.at/en/order/price-list.html

Fertilisation

Soil fertilisation is partly based on the spreading of compost and possibly of biochar. If these amendments can be purchased, they can also be products on the farm. Equipment (oven for biochar...) and specific tools can prove necessary to produce them and spread them.

TOOLS	AVERAGE PRICE	MINIMUM PRICE	MAXIMUM PRICE	NOTES
Compost sieve	€30 incl. VAT	€16 incl. VAT	€100 incl. VAT	
Thermometer				Compost thermometer (for monitoring the temperature of warm layers) From -10 to 90°C. 44cm length
Biochar oven	Self-fabrication possible[9]			
Buckets	€6 per 16L bucket			

Seeding

TOOLS		AVERAGE PRICE	MIN. PRICE	MAX. PRICE	NOTES
Direct sowing	6-row seeder (Coleman)[10]	$625 (€540 approx.)			
	4-row seeder	$11,249 (€215 approx.)[11]			
	Seed bed roller	€50	€30	€85	
Container sowing	Sowing trays	€3 per tray (sold in a box of at least 25 trays). Three different sizes (104, 96, 60).			Three different sizes (104, 96, 60), Herkuplast[12]
	Herrissons	€20 per tray			Three size moulds (104, 96, 60)
	Individual pots	€3.50 per pack of 100			10 × 10 × 10
	Seed table	Self-fabrication possible Think about ergonomics; seeding table height adapted to the size of the market gardener			
Seed cabinet		Recovery or self-fabrication possible Cool, dry cabinet. Pay attention to germination rate and seed conservation			

Transplant

TOOLS	AVERAGE PRICE	MIN. PRICE	MAX. PRICE	NOTES
Dibber	Self-made			
Dibber	€6	€4	€35-40 incl. VAT (bronze planters from PKS)	Self-fabrication of graduated T-shaped planters every 10cm or every 5cm
Dibber template	Self-made Spacing: Every 9, 11, 15cm, staggered			
Transplanter or garden trowel	€10	€6	€36 incl. VAT (trowels in bronze from PKS)	

Irrigation

TOOLS		AVERAGE PRICE	MIN. PRICE	MAX. PRICE	NOTES
Watering gun or lance		€40	€15	€60	Plan for one watering gun/lance
Watering cans		€6			Plan for four watering cans
Sprinkler		€4	€2 incl. tax, i.e. €1.67 excl. tax	€5 inc. tax, i.e. €4.17 excl. tax	Plan for two sprinklers
Pipes		€122 for 50m			Two of 25m
Polytunnel	Drip line	€44			Two drip lines per growing bed (bed 20m long, therefore 40m per board) over 400m²
	Fixed sprayer	€1,100			Four lines at 20m
Exterior	Drip line	€66 Streamline reel, 320 µ sheath, 780m			Two drip lines per growing bed (beds 20m long, therefore 40m per bed) over 600m²
Metal staples		€12 for 100 pc			Have a stock of 200 metal staples. Provide a dedicated storage location protected from humidity.

Point of vigilance: integrate a water management system according to the context and configuration of the farm. Different options are possible: drilling, meter connected to network water, electric pump... At Bec Hellouin Farm, the water management system is based on an electric pump (flow rate 7.2 cubic metres per hour) with a 100 litre tank (pump: 2,500 euros, 100 litre tank tank: 1,000 euros, small associated equipment: 1,000 euros).

7. More information on self-making a biochar oven on the Fabriculture site (Vincent Legris): www.lafabriculture.fr/Four-a-BioChar.html.
8. More information on the Coleman seeder (seeder 6-row precision) on the Johnny's Seeds website: www.johnnyseeds.com/p-6613-six-row-seeder.aspx.
9. More information on the 4-row seeder on the Johnny's Seeds website: www.johnnyseeds.com/p-7873-four-row-pinpoint-seeder.aspx.
10. More information on seed plates and on the unmolding plates on the company website Herkuplast: www.herkuplast.com/fr/accueil.html.

Weeding and Mulching

TOOLS	AVERAGE PRICE	MIN. PRICE	MAX. PRICE	NOTES
Manual weed harrow[13]	$33.95 (€29 approx.)			Weeds at the young seedling stage
Micro-hoes, precision wire weeders[14]	€40 (€34.50 approx.)	€5 incl. VAT	€120 incl. VAT (oscillating rack in bronze from PKS)	Choose from three different sizes for different levels of precision
Short-handled claw	€10	€5 incl. VAT		
Weeding knife	€10	€5.50 incl. VAT		
Pyroweeder burner	€140 (backpack or wheeled one nozzle)		€270 (backpack or wheeled, four nozzles)	There are pyroweeders on wheeled carts as well as mixed pyroweeders (backpack and wheels) The number of nozzles is also a variable
Heavy tarp				
Row cover				
Straw cutter				
Meat cleaver				

Crop Management

The management of certain crops may require specific material concerning:

- Protection against the cold: shelters, sails forcing type P17, Nantes tunnels...
- Staking on a stake (made of bamboo for aubergines, peppers and melons or string for tomatoes and cucumbers) or on netting (peas and beans rowing, for example).

TOOLS		AVERAGE PRICE	NOTES
Shelter/polytunnel		Used price: €6,500-8,000 New price: €12,500-15,000	Price given approximately 600m² Price includes metal structure and plastic cover
P17	Row cover	€125 for two rolls of 250m Lutrasil Thermoselect 2 × 250m, 19g	Provide as many pieces as the number of beds (length of each bed 20m)
	Hoops	Metal	Provide around 10 hoops per 20m bed
Metal staples		€12 for 100	Have a stock of 200 metal staples Provide around 20 staples per 20m bed Provide a dedicated storage location, protected from humidity
Row cover low tunnels	Plastic tarp	€280 for two rolls of 220m, Therm Cristal, 3 × 220m, 80 μ	Provide enough row cover for four beds
	Stakes (bamboo)	€2.10 each Galvanised stakes Ø 6, two clasps, length: 3.22m	Provide around 10 hoops per 20m bed
Rollers		€155 for box of 350	Tomato roller hooks, 32m of wire Plan for at least 1,000
Clips		€55 for 10,000	Tomato clips 23mm in diameter
Stakes (bamboo)		€35-50 for 100 (210cm)	600 matched according to desired size
Roll of disposable netting		€30 for 100m roll (2m wide)	

Shelter/polytunnel: the price indicated in the table corresponds to an entry-level shelter model. Several options are possible but are accompanied by a increase in the amount of investments to be planned.
- Opening systems for aeration and ventilation of the polytunnel: number and methods. Favour winding systems, especially when working alone.
- Straight-edged polytunnels save space and provide greater working comfort.

Crop Protection and Care

TOOLS	AVERAGE PRICE	MIN. PRICE	MAX. PRICE	NOTES
Compost tea system	€20	€5	€50	Possibility of using an aquarium pump or diffuser
Manual backpack sprayer	€120			
Sieve to filter manure	€20			Fine mesh
Barrel	€50	€30	€70	6-120 L
PPE (personal protective equipment) to standards	€150 for purchase, then up to €50 for renewals (cartridges, gloves and overalls if not washable)			Complete set of protective clothing: A2P3 mask, nitrile gloves, protective glasses and suit (chemical protection cat. 3)
Phytosanitary cabinet with keys	€500	€400	€600	

Areas for improvement and options:
- Insect nets (270 euros per 250 metre roll) for the fight against insects (mechanical barrier) and the quality of vegetables: provide several nets and one set of staples to secure them.
- An electric sprayer (400 euros) with a 20 litre tank.

11. More information on the manual weeder harrow on the Johnny's Seeds website: www.johnnyseeds.com/p-9789-tine-weeding-rake-21-wide.aspx.
12. More information on micro-hoes and weeders precision on the Johnny's Seeds website: www.johnnyseeds.com/p-6691-narrow-collinear-hoe-3-fixed-blade.aspx, www.johnnyseeds.com/p-5599-long-handled-wire-weeder.aspx, www.johnnyseeds.com/p-5599-long-handled-wire-weeder.aspx

Harvesting, Washing, Storing

TOOLS	AVERAGE PRICE	MIN. PRICE	MAX. PRICE	NOTES
Scale	€150	€30	€430	Accuracy: 1g or 5g depending on model Maximum weight: 30kg or more depending on model
Wheelbarrow	€340			Wheelbarow (1 wheel)
Stapler	€20	€10	€35	Plus replacement staples
Paper, plastics, towels to control humidity	€6 per roll, sold in packs of six			Paper towels 150m rolls
Potato nets	€20-25 for 100			The price varies depending on the container For market gardening, nets of 20-25kg capacity are sufficient
Watering can	€5	€3	€10	
Bucket	€6 for 16 L bucket			
Harvest trays	€6 each			Stackable and foldable plastic crates for harvesting Dimensions: 58 × 35 × 12cm
Galvanised harvest baskets	€15 for 15 L basket €18 for 30 L basket			Use baskets of different sizes (three sizes) The price depends on the volume of the basket. Possibility of finding these baskets second-hand.
Crates	€25 for three			Stackable and foldable plastic crates for harvesting Dimensions: 50 × 30 × 26.5cm
Crates	€0.80 each	€0.35 each, used	€1 each new	If possible, collect crates from customers weekly. Be careful to store crates in a dry place, protected from humidity.
Plastic boxes	€10 for 50			Plastic boxes with lid for 'freshness'. Capacity: 1kg
Elastic bands	€2 per box of 230 bands sold in sets of 5 boxes (1,200 × 1.5mm). €8 per box of 420 elastics (1,200 × 10mm).			When possible, collect bands from customers weekly

Areas for improvement and options:
- Access to running water;
- A mesclun (young salad leaves) harvester;
- A wringer;
- An ergonomic and practical washing station (to be adapted to match the height of the market gardener);
- A harvest cart or wheelbarrow.

Storage

Having suitable storage locations is essential for the proper conservation of vegetables, with optimal temperature, humidity and light. The needs of storage extremely variable depending on the system and location of production, but also the method of marketing. It is not going to be a straightforward estimate of the investments needed. However, it is entirely possible to list potential needs:

- Harvest shelves, baskets, boxes, crates and nets.
- A dry frost-free area for squash storage.
- A frost-free zone such as a cellar.[15]
- A drying area.
- Shelves for drying (aromatics, garlic, onions, shallots, etc.).

Sales and Marketing

Sales needs will vary depending on a farm's production and existing systems and outlets. The table below is presented for informational purposes only. It lists the equipment used at Bec Hellouin Farm for production and sales.

The equipment must:

- Display prices.
- Weigh and record purchases.
- Package for retail and wholesale.

13. For more information on building a natural cellar, read: Claudia Lorenz-Ladener, *Building a Natural Cellar: Construction and development of spaces for the conservation of fruits and vegetables*, Ulmer, Paris, 2012.

TOOLS	AVERAGE PRICE	MIN. PRICE	MAX. PRICE	NOTES
Signs	€20 for 10 blank slate pieces			Collect a batch of old slate from a flea market
Celloplast white felt	€4.50 each			Chalk marker medium bullet point (1.8-2.5mm). Use washable inks.
Scale	€150	€30	€430	Accuracy: 1g or 5g depending on model Maximum weight: 30kg or more depending on models
Calculator	€5	€1	€25	
Galvanised harvest container	€15 for 15 L container €18 for 30 L container			Use different sizes (three sizes) The price depends on the volume of the basket. Possibility of finding these in flea markets or second-hand.
Crates	€25 for three			Stackable and foldable plastic crates for harvesting Dimensions: 50 × 30 × 26.5cm
Wicker baskets	Possibility of finding these baskets in flea markets or second-hand			
Crates	€0.80 each	€0.35 each, used	€1 each, new	If possible, collect crates from customers weekly
Cardboard trays	125g trays: €109 per batch of 1,300 250g trays: €90 per batch of 800 500g trays: €75 per batch of 500			
Paper bags	€15 for 1,000 (capacity 1,000g) €20 for 1,000 (capacity 2,000g)			Brown paper bags, food grade, for packaging for sale of fruits and vegetables. If possible, collect paper bags from customers weekly.
Bag sealer	€19			Plus adhesive for sealer: €10 for six rolls of 66m Plus plastic bags

Optional area for improvement: a cash register.

Small Tools

TOOLS	AVERAGE PRICE	MIN. PRICE	MAX. PRICE	NOTES
Scissors	€20	€2	€25	
Knife	€10	€8	€15	
Pruners	€40	€20	€60	Japanese pruners: Japanese tools are of very high quality, with hand-forged, hard and very sharp steel. Their cost is high (from €90-110) but they are effective and very sturdy.
Shears	€40	€20	€60	Japanese shears from €110-130
Serrated knife	€11	€10	€20	Serrated greens knife from Johnny's Seeds: $10.60 (€9 approx.)[16]
Pruning hooks, sickles	€40	€13	From €60-100 (bronze tools from PKS)	
Cleaver	Possibility of finding this tool in flea markets or second-hand			
Sickle	€150 (blade + handle)			
Scythe	€160 (blade + handle)			
Sharpening stone	€15	€10	€20	
Saw	€30	€20	€40	
Hatchet	€30	€8	€47	
Axe	€50			

Work the Soil, Decompact, Aerate, Weed, Mulch, Dig

TOOLS		AVERAGE PRICE	MIN. PRICE	MAX. PRICE	NOTES
Hoe	large	€20	€7	€69 (large bronze garden hoe and cultivator combo from PKS)	
	small	€10	€5	€50 (small bronze hoe and rake combo from PKS)	
Rake		Classic rake 14 teeth, straight or curved: €17	€4	€90 (bronze rake from PKS)	Preparation of flat beds and rounded beds
Fork		€30	€20	€40	
Pitchfork		€30	€20		
Rock rake		€50	€20	€80	
Four prong cultivator		€35	€16	€135 (bronze hook from PKS)	
Spade		€60	€18	€150 (bronze spade from PKS)	
Pick axe		€30	€20	€40	
Shovel		€30	€10		
Crowbar		€50	€30	€90	
Hoe		€20			

Tie, Bind

TOOLS	AVERAGE PRICE	NOTES
Elastic bands	€2 box of 230 sold in five boxes (1,200 × 1.5mm) €8 box of 420 bands (1,200 × 10mm)	If possible, collect weekly from customers
Wire spools	€4.40 for 100m roll	
Rope	€32 for 30m	Diameter: 20mm
Cord	€2.50 for 20m	
String	€20 for 330m	
Raffia	€15 for 1kg box	Natural raffia
Pegs	€1.50 for 24	Wooden or stainless steel

Transport

TOOLS	AVERAGE PRICE	MIN. PRICE	MAX. PRICE	NOTES
Galvanised harvesting container	€15 for 15 L container €18 for 30 L container			Different size containers (three sizes) The price depends on the capacity of the container. Possibility of finding these in flea markets or second-hand.
Buckets	€6 for three at 16 L			Plastic buckets
Harvesting racks	€6 for three			Stackable and foldable plastic racks for harvesting Dimensions: 50 × 30 × 26.5cm
Crates	€25 for three			Stackable and foldable plastic crates for harvesting Dimensions: 50 × 30 × 26.5cm
Boxes	€0.80 each	€0.35 each, used	€1 each, new	If possible, collect weekly from customers Be careful to store in a dry place, protected from humidity
Wheelbarrow	€90	€35	€260	
Wheelbarrow-workshop	Can be self-made from a wheelbarrow			The wheelbarrow-workshop allows you to move around the gardens with all necessary tools

Options for improvement:
- A multifunctional, versatile tool trolley: transport tools, seedling trays, plants, compost, hoes... adapted to the size of pathways and beds. Choose a cart with ample capacity, but pay attention to pathway width.
- Chinese wheelbarrows (see the work of Dominique Soltner on this subject): self-manufacturing possible.[16]

Write, Note, Identify

TOOLS
Label markers
Crop labels
Crayons
Chalk
White crayon
Calendar

Clean and Maintain

TOOLS	AVERAGE PRICE	MIN. PRICE	MAX. PRICE	NOTES
Brooms/brushes	€15			32cm head with handle
Broom	€19			80cm head with handle
	€16			Rice broom
Brush	€5			
Nylon brush	€5			Provide a dedicated storage location near a water point or washing station
Nail brush	€2	€1	€4	
Sponge	€1 each			

14. More information on the serrated knife on the Johnny's Seeds website: www.johnnyseeds.com/p-7854-serrated-greens-knife-6-blade.aspx.

Enclosures

TOOLS	AVERAGE PRICE	MIN. PRICE	MAX. PRICE	NOTES
Maul	€40	€20	€50	
Stakes	€3.50 each			Depending on the uses: metal, wood (use rot-proof wood, if possible untreated) or plastic (electric fences).
Net for chickens	€160-190			Chicken netting, 32m: door, ropes and double-pointed stakes included. Add staples and concrete bars during installation.

Miscellaneous

TOOLS	AVERAGE PRICE	MIN. PRICE	MAX. PRICE	NOTES
Toolbox	€350	€125-150	€830-1,000	Essential tools: screwdriver, hammer, wrench, open-end wrench, pipe wrench, hacksaw, level, square, drill bits, nails, screws, pliers and multi-purpose pliers...
Tape measure	€5.50-6.50			5m length
Funnel	€10	€5	€25	
Torch	€50			
Thermometer	€12.50			Compost thermometer (for monitoring the temp of hot layers) From -10 to 90°C. 44cm long
Small lightweight stepladder	€90	€50-60	€80-100	
Drill and spare battery	€170-200 Battery voltage: 18 V plus a spare battery. Pay attention to the amperage.			Using the Coleman microcultivator

Taking Care of the People

It's also important to invest in protective equipment and clothing appropriate for market gardening activities.

TOOLS	AVERAGE PRICE	MIN. PRICE	MAX. PRICE	NOTES
Gloves	€10	€2	€20	Comfortable gloves for light market gardening work such as weeding
Personal protective equipment up to regulations	€150 plus up to €50 for renewals (cartridges, gloves and overalls if not washable)			Complete protective clothing: A2P3 mask, nitrile gloves, protective glasses and suit (chemical protection cat. 3)
Knee pads	€60	€22		There are also weeding stools, memory foam knee protector cushions
First aid kit	€15			To be checked regularly and supplemented or renewed as necessary

Don't forget :
- Suitable shoes (safety, warm, waterproof and comfortable shoes);
- Suitable clothing (warm, waterproof, comfortable);
- Coffee maker and thermos;
- Earplugs, earmuffs, helmet and protective visor for cutting work and/or earmuffs and visor for brushcutting work will also be required.

Biodiversity

Permaculture design must promote the presence of beneficiaries involved in crop protection. For this, it is necessary for the beneficiaries to be able to find their shelter, their cover and reproduction. It's also necessary to protect yourself against any harm they could cause.

TOOLS	AVERAGE PRICE	NOTES
Nest boxes for nocturnal and diurnal animals	Do it yourself for between €5-170	See LPO catalogue for advice and wide choice Use nest boxes with a 27mm maximum opening, to encourage the arrival of the most insectivorous and territorial birds and fewer granivores
Feeders	Make it yourself for €4-50	See LPO catalogue for advice and wide choice Consider integrating permanent water features (summer and winter)
Rat trap	€30	

SUMMARY TABLES – INVESTMENTS TO CONSIDER FOR TOOLS

Below we present two tool lists. They will be useful in determining what you will need and the investments to consider.

List 1 presents a restricted range which nevertheless allows you to start a market gardening activity well. In this list, there is no greenhouse; it is replaced by tunnels. It is suitable for market gardeners who are starting their business with limited investment.

List 2 (which includes the first) is more extensive, and includes all the tools we use daily on our farm, 11 years in, with several permanent market gardeners. This list is therefore suitable for a farm that produces all year round in good conditions.

List 1: A limited range that allows you to start a market gardening activity in good conditions

	TOOL	UNIT PRICE BEFORE TAX (€)	QUANTITY	TOTAL PRICE BEFORE TAX (€)
SOIL PREPARATION	Campagnole	275	1	275
	Broadfork	75	1	75
	Classic rake	17	2	34
	Grass rake	41	1	41
	Hoe	63	1	63
	Rope	DIY		
Total				**488**
FERTILISATION	Biochar oven	DIY or buy biochar		
	Buckets	6	5	30
Total				**30**
SEEDING	6-row seeder	565	1	565
	Roller	50	1	50
	Labels (box of 25 labels at €3 each)	75	8	600
	Individual cups (100)	3.5	30	105
	Seeding table	DIY		-
	Seed cabinet	DIY		
Total				**1,320**
HARVEST	Planter with guard	Self-made		
	T-shaped planter	Self-made		
	Classic planter	6	2	12
	Transplanting template plate	Self-made		
	Transplanter	10	2	20
Total				**32**
IRRIGATION	Watering gun or lance	40	1	40
	Watering cans	6	4	24
	Watering rose	4	2	8
	Pipes (50m)	122	1	122
	Metal staples (100)	12	2	24
Total				**218**
WEEDING AND MULCHING	Manual weeder harrow	30	1	30
	Precision weeders	36	3	108
	Short-handled claw	10	2	20
	Weeding knife	10	2	20
	Thick silage-type tarpaulin	100	1	100
	Chopper	10	1	10
Total				**288**

	TOOL	UNIT PRICE (€) BEFORE TAX	QUANTITY	TOTAL PRICE BEFORE TAX
CROP SUPPORTS	P17 (roll of 250m)	62.5	2	125
	Tarp for polytunnel (roll of 220m)	190	1	190
	Arches for polytunnel	2.1	100	210
	Bamboo stakes (100)	40	6	240
	Netting (100m)	30	1	30
Total				795
CROP PROTECTION AND CARE	Manual backpack sprayer	120	1	120
	Sieve to filter manure	20	1	20
	Blue plastic barrel	50	6	300
	Standard protective gear	150	1	150
	Phytosanitary cabinet	100	1	100
Total				690
SMALL TOOLS	Scissors	20	2	40
	Opinel knife	10	2	20
	Secateurs	40	2	80
	Shears	40	1	40
	Toothed knife	11	2	22
	Pruning hooks, sickles, crescents	40	2	80
	Sharpening stone	15	2	30
	Saw	30	1	30
	Hatchet	30	1	30
Total				372
WORK THE SOIL, DECOMPACT, AERATE, WEED, MULCH, DIG	Big hoe	20	1	20
	Small hoe	10	2	20
	Fork	30	2	60
	Spade fork	30	1	30
	Rake	35	1	35
	Spade	60	1	60
	Pickaxe	60	1	60
	Shovel	30	2	60
	Crowbar	50	1	50
	Hoe	20	1	20
Total				415
ATTACH, BIND	Wire	4.4	1	4.4
	Rope (30m)	32	1	32
	Cord (20m)	2.5	1	2.5
	String (330m)	20	1	20
	Raffia (1kg)	15	1	15
	Pegs (24)	1.5	10	15
Total				89
CARRYING	Wheelbarrow	90	2	180
	Wheelbarrow-workshop	Self-made		
Total				180
WRITE, NOTE, IDENTIFY	Labels to plant (100)	8	2	16
	Chalk markers (four)	17	1	17
	Markers	0.9	1	0.9
	Pencil	3	1	3
	Diary	23	1	23
Total				60
CLEAN, MAINTAIN	Broom	15	1	15
	Dustpan and brush	5	1	5
	Nylon brush	4	1	4
	Fingernail brush	2	1	2
	Sponge	1	2	2
Total				28
MISCELLANEOUS	Tool box	350	1	350
	Measuring stick (5m)	5.5	1	5.50
	Funnel	10	1	10
	Torch	50	1	50
	Small lightweight stepladder	90	1	90
Total				505.50
Total price before tax				5,510.50
Grand total				6,612.60

List 2: The tools we use on a daily basis at Bec Hellouin Farm, in our eleventh year

	TOOL	UNIT PRICE BEFORE TAX (€)	QUANTITY	TOTAL PRICE BEFORE TAX (€)
PREPARING THE SOIL	Campagnole	275	1	275
	Microcultivator	450	1	450
	Broadfork	75	1	75
	Classic rake	17	2	34
	Rake 80cm	77	1	77
	Lawn rake	41	1	41
	Hoe PKS	63	1	63
	String	Self-made		
Total				1,015
FERTILISATION	Compost sieve	25	1	25
	Thermometer	13	2	26
	Oven for biochar	Self-made		
	Buckets	6	5	30
Total				81
SEEDLINGS	Seeder for six	565	1	565
	Seeder for four	225	1	225
	Roller	50	1	50
	Seedling labels (box of 25 at €3)	75	12	900
	Unmoulding plate	20	3	60
	Pots (100)	3.50	30	105
	Seed table	Self-made		
	Seed cabinet	Self-made		
Total				1,905
TRANSPLANTING	Planter with guard	Self-made		
	T-shaped planter	Self-made		
	Classic planter	6	2	12
	Transplanting template	Self-made		
	Transplanter	10	2	20
Total				32
IRRIGATION	Gun or lance	40	1	40
	Watering cans	6	4	24
	Watering rose	4	2	8
	Pipes (50m)	122	1	122
	Drip hose	44	1	44
	Greenhouse sprinkler	275	4	1,100
	Exterior drip hose	66	1	66
	Metal staples (100)	12	2	24
	Please note: a watering system must be adapted according to contexts and resources (drilling, electric pump, agricultural meter, etc.)			
Total				1,428
WEEDING AND MULCHING	Manual weed harrow	30	1	30
	Precision weeders	36	3	108
	Short-handled claw	10	2	20
	Weeding knife	10	2	20
	Knee pads	60	2	120
	Backpack pyroweeder	140	1	140
	Thick silage-type tarpaulin	100	1	100
	Woven canvas tarpaulin	130	3	390
	Straw chopper	250	1	250
	Chopper	10	1	10
Total				1,188
CROP MANAGEMENT	Second-hand greenhouse (600m²)	6,700	1	6,700
	P17 (250m roll)	62.50	2	125
	Tarpaulin for polytunnel (220m roll)	190	1	190
	Arches for polytunnel	2.10	100	210
	Winders (box of 350)	155	4	620
	Clips (set of 10 000)	55	1	55
	Bamboo stakes (100)	40	6	240
	Netting (100m)	30	1	30
Total				8,170

	TOOL	UNIT PRICE BEFORE TAX (€)	QUANTITY	TOTAL PRICE BEFORE TAX (€)
CROP PROTECTION AND CARE	Bubbler for compost tea	20	1	20
	Insect netting	270	1	270
	Manual backpack sprayer	120	1	120
	Electric sprayer	400	1	400
	Sieve to filter manure	20	1	20
	Blue plastic drum	50	3	150
	Regulation safety gear	150	1	150
	Phytosanitary cabinet	500	1	500
Total				1630
HAND TOOLS	Scissors	20	2	40
	Opinel knife	10	2	20
	Secateurs	40	2	80
	Shears	40	1	40
	Toothed knife	11	2	22
	Prune hooks, sickles, crescents	40	2	80
	Machete	150	1	150
	Fork	160	1	160
	Sharpening stone	15	2	30
	Saw	30	1	30
	Hatchet	30	1	30
	Chopper	50	1	50
Total				732
WORK THE SOIL, DECOMPACT, AERATE, WEED, MULCH, DIG	Mattock	20	1	20
	Small mattock	10	2	20
	Fork	30	2	60
	Spade fork	30	1	30
	Potato fork	50	1	50
	Rake	35	1	35
	Spade	60	1	60
	Pickaxe	60	1	60
	Shovel	30	2	60
	Crowbar	50	1	50
	Hoe	20	1	20
Total				465
ATTACH, BIND	Wire	4.40	1	4.40
	Rope (30m)	32	1	32
	Clothesline (20m)	2.50	1	2.50
	String (330m)	20	1	20
	Raffia (1kg)	15	1	15
	Pegs	1.50	10	15
Total				89
TRANSPORT	Wheelbarrow	90	2	180
	Large wheelbarrow	340	1	340
	Wheelbarrow-workshop	Self-made		
Total				520
WRITE, LABEL, IDENTIFY	Plant labels (100)	8	2	16
	Chalk (box of four)	17	1	17
	Markers	0.9	1	0.90
	Crayons	3	1	3
	Calendar	23	1	23
Total				60
CLEAN, MAINTAIN	Broomstick	15	1	15
	Broom	16	1	16
	Dustpan and brush	5	1	5
	Nylon brush	4	1	4
	Nail brush	2	1	2
	Sponge	1	2	2
Total				44
MISCELLANEOUS	Tool box	350	1	350
	Measuring stick (5m)	5.5	1	5.50
	Funnel	10	1	10
	Torch	50	1	50
	Small lightweight stepladder	90	1	90
	Battery-powered drill	170	1	170
Total				675.50
Total price before tax				18,034.50
Grand total				21,641.40

ACKNOWLEDGEMENTS

Writing this manual has been a long journey, a great shared adventure! We wouldn't have been able to do it without the commitment of many people by our side. We cannot cite all those who have contributed to this project in one way or another here, but know that you are all in our hearts and that our gratitude is infinite!

Our thanks go first and foremost to our families, our parents and especially our wonderful girls: Lila, Rose, Shanti and Fénoua. Thank you for your understanding and patience and support on all of those exhausting days! Your love carries us unceasingly. We hope that this manual will help, if only a little, to leave you and all the children of the world with a beautiful planet!

It took a lot of courage to agree to publish an 'endless' manual like this one, the completion of which was postponed from year to year, as the number of pages increased. Thanks to our dear editors who believed in it and have always supported us with so much enthusiasm. Thanks to Anne-Sylvie Bameule, Jean-Paul Capitani and Françoise Nyssen for their committed and visionary publishing, their immense talent, big hearts, and precious friendships.

Thanks to Camille Joyeux, scientific and technical content manager of this manual, for her exceptional commitment, working with us for two years. Camille, your hard work, the seriousness that you contributed to the drafting of the summary notes, the realisation of tables, proofreading, and your helpful advice contributed immensely to making this manual what it is.

Thanks to Philippe Laborde, the designer who supported all of our books for more than fifteen years. Philippe, I could not tell you how much these years spent together designing the model were made pleasant by your intense listening, your encouragement, your creativity and your taste for a job well done!

We would like to pay tribute to Aïté Bresson, the best of editors! Aïté, you took charge of the manual when it was still a huge disparate puzzle and with infinite patience, assembled the pieces and corrected everything that needed editing, very effectively supported by Lauranne Valette with the ultimate proofreading. We can't tell you how much your experience and kind words have brought us! Our thanks also go to the whole team at Actes Sud, in particular Isabelle Alliel, Karima Chabaud, Émeline Lacombe, Mencia Morere.

This book is wonderfully illustrated by the photos by Claudius Thiriet, farmer and photographer, friend for nearly twenty-five years. Claudius, after having crisscrossed the oceans together aboard *Fleur de Lampaul*, I'm glad the earth gave us the opportunity for a new fruitful collaboration! A big thank you also to other photographers who contributed to this manual: Antoine Agoudjian, Babeth Aloy, Emmanuel de Bettignies, Carlo Bettinelli, Franck Boucourt, Gauthier Chapel, Laure Chavanne, Gilles Degroote, Xavier Dubois, Martin Egbert, Lauren Fleishman, Nicolas Guiraud, Fruzsina Gyertyan, Pascal Hanrion, Christophe Jammet, Christine Kristof, Armand Laskowski, Thierry Mesnard, Max Riché, Andrea Sham, Julien Vanlerberghe, Nicolas Vereecken, Pauline de Voghel.

This manual has benefited from the generous support of citizens of the world with a big heart and we want to thank our friends and fellow travellers Charlotte de Mévius, François and Françoise Lemarchand and Michaël Lunt for their commitment to us and their trust in us.

The farm team is a valiant crew, which holds its course against all odds! We express our gratitude to those who brought their talents, their humour and their joy to this adventure, yesterday or today: Jean-Claude Bellencontre, Jean-Pierre Bellencontre, Hugo-Martial Bonnet, Justin Bourel, Laure Chavanne, Thomas Henriot, Christine Joly, Yohann Jourdan, Anne-Noëlle Jubelin, Édith Legay, Fanny Marquer, Luigi Morin, Isabelle Perron, Fabien Prudhomme, Baptiste Saulnier, Frédéric Sauvadet and Cécile Thibault.

The contents of this manual owe much to the members of the scientific committee of the farm and to the researchers who conducted studies on our site – however, let us specify that the opinions expressed in these pages are our own! We warmly thank Stéphane Bellon, agronomist, INRA; Lydia and Claude Bourguignon, agronomists, co-founders of LAMS (Laboratory of Soil Microbiology Analysis); Alain Canet, French Association of Agroforestry; Clara Carrayrou, agronomist; Gauthier Chapelle, biomimetician, Greenloop; Jean-Thomas Cornélis, soil scientist, Gembloux University Agro-Bio Tech (Belgium); Nathalie Corroyer, agronomist, Regional Chamber of Agriculture of Normandy; Martin Crawford, Agroforestry Research Trust

(England); Philippe Desbrosses, agroecologist, Ferme de Sainte-Marthe; Jean-Marie Deshoux, Pur Projet; Luc Devaux, agronomist; Marc Dufumier, agronomist, AgroParisTech; Paul Emptaz, agronomist; Louise Gehin, agroecologist; Morgane Goirand, agronomist; Alain Grandjean, Carbone 4; Sacha Guégan, Nature Permaculture; François Léger, umr sad-apt, AgroParisTech/INRA; Alexis de Lidekercke, agronomist, Froidefontaine farm; Kevin Morel, agronomist, AgroParisTech; Isabelle Poirette, Seeds movement of life; Arthur Rouanet, Pur Projet; Pierre Scheercousse, Tree & Landscape 32; Pablo Servigne, researcher, writer; Eskani Siruguet, agroecologist; Stefan Sobkowiack, arborist, founder of Miracle Farms (Quebec); Valentin Sohy, agronomist; Dominique Soltner, agronomist and author; Pierre Stassart, agronomist; Serge Valet, agronomist; Nicolas Vereecken, agronomist, ecologist; Margaux Villebrun, bioengineer; François Warlop, Organic Farming Research Group (GRAB).

A huge thank you to the partners of the Sylva Institute, whose trust and loyal commitment make it possible to lead innovative programmes: Akuo Foundation, Ardian Foundation, Daniel and Nina Carasso Foundation, Fondation de France, Fondation Lemarchand for the balance between Man and the Earth, Foundation Lunt, Picard Foundation, Terra Symbiosis Foundation, Paris City Hall, without forgetting our friends and patrons Gilles Ghesquières, Agnès Martin, Thierry Plojoux, Nicolas and Marina Plowiecki (balt-newco). Each of these foundations is led by a great team whom we have had a lot of fun working with.

Our gratitude is enormous for our friends at the Sylva Institute, starting with its honorary president, Philippe Desbrosses, its president Lucile de Cossé Brissac, as well as Louis-Albert de Broglie, Laurent de Chérisay, Jean-Baptiste Dumont, Sébastien Henry, François Lemarchand, Alexandre Poussin and Frédéric Proniewski.

Volunteers and students who have completed internships on the farm are numerous, and we express our gratitude to each of them. Thank you for working together and for the good times that we shared.

Among the many people and friends who have supported and encouraged us during this adventure, we would like to give special thanks to Rémi Algis, Bernard Alonso, Marie-Marie Andrasch, Steve Arcelin, Claude Aubert, Laurent Auguste, Huguette Autin, Florence Ballay, Thibault Basquin, Christophe Beckaert, Armand Bongrain, Antonin Bonnet, Cybèle de Brem, Françoise Brenckman, Arnaud Brulaire, Chloé Bulgheroni, Eliot Coleman, Loïc Coutelle, Arnaud Daguin, Stéphanie Deflandre, Pascal Depienne, Sylvie Dévigne, Cyril Dion, Jérôme Doncieux, Philippe Dubarry, Gérard Dufils, Bertrand and Béatrice Finet, brother Raphaël Flaujac, Jean-Martin Fortier, Jean-Yves Fromonot, Training Sainte-Marthe; Ideline Genette, Thierry Gissinger, Thierry and Pascale Glaizot, Marc Grollimund, Jérôme Henry, Bruno and Martine Hermenault, Jean-Olivier Héron (†), Bertrand Hervieu, Jean-Pierre and Blanche Hopsore, Jean-Marie Hullot, Aymeric Jung, Patricia Jung-Singh, Annie Kerangueven, Julie Laffond, Minnie and Romain Lassus, Mélanie Laurent, Tristan Lecomte, Jean-Jacques Legarrec, Vincent Legris, Antoine and Agnès Lemarchand, Philippe Lemarchand, Hélène Le Teno, Martine, Joseph and Syméon Losfeld, Dorothée Machils, Blanche Magarinos-Rey, Ludivine Mesnard, Antoine Minkowski, Alexandra Mitsotsaki, Lionel Mourlin and his team, Éric Mulliez, Emmanuel Oblin, Serge and Carina Orru, Jean-Marie Pelt (†), Roger Pillot, Ian Pinault, Jérôme Pousset, Mona Puill-Stephan, Pierre Rabhi, Katinka Rambert, Jean-François Renard, Hélène Rollin, Éric Scotto, Stéphane de Swaef, Claude Taleb, Alessandro Thellung, Françoise Vernet and Mona Zeutschel.

Charles especially thanks Laure Lemarchand and Isabelle Minne, with whom he led the Indoor Ecology training course, effectively supported by Vanessa Correa, Élodie Loquet, Laura Mentasti and Élise Ninnin.

We warmly thank the municipal team of Bec Hellouin, who has always supported our projects, particularly Jean-Paul Vittecoq and Pascal Finet. Our gratitude also goes to our brothers and sisters of the monastic communities of Bec Hellouin.

La Ferme du Bec Hellouin benefited from the support of the Haute-Normandie region, from the Eure General Council, from the Eure Chamber of Agriculture, the BPI, GRAB of Haute-Normandie. Thank you for your confidence. We are also grateful to those who work to develop permaculture in France and around the world; they are growing in number every year. A special thanks to the team at L'Université Populaire de Permaculture, and Brin de Paille.

BY THE SAME AUTHORS

Perrine and Charles Hervé-Gruyer
Permaculture. Guérir la Terre, nourrir les hommes, Actes Sud, 2014.
(Translated into English as *Miraculous Abundance: One Quarter Acre, Two French Farmers, and Enough Food to Feed the World*, Chelsea Green Publishing Co., 2016.)

Perrine Hervé-Gruyer
La Relaxation en Famille, Presses de la Renaissance, 2008.

Charles Hervé-Gruyer
Sagesse d'ailleurs pour vivre aujourd'hui (ouvrage collectif, Federika Van Ingen), Éditions des Arènes, 2016.
L'avenir est en nous (ouvrage collectif), Dangles, 2014.
La Femme Feuille (roman), Albin Michel, 2007.
Cunas, les Indiens du Corail, Gallimard Jeunesse, 1994.
Wayanas, les Indiens du Fleuve, Gallimard Jeunesse, 1994.
Bijogos, les Grands Hommes de l'archipel, Gallimard Jeunesse, 1993.
Imragen, les pêcheurs du désert, Gallimard Jeunesse, 1992.
Les Enfants Dauphins, Gallimard Jeunesse, 1990 (Médaille de l'Académie de marine).

PHOTO CREDITS

All photos © Charles Hervé-Gruyer, except for:
© Pascal Hanrion: p.2;
© Claudius Thiriet: pp.6, 29 b., 32, 36, 38 m. and b., 39, 41 b., 43 m., 44 t., 46, 48, 65 t., 66 b., 70, 73 b., 74, 108, 109, 110 b., 111 t., 119 t., 123 t., 124 t., 125 t., 128, 137, 142, 145 t., 146 t., 147, 156, 158, 159, 160 t., 164, 169, 177 t., 183 m.-b., 187, 198 to 200, 203, 207 m., 209 b., 218 b., 219 t., 229, 232 to 238, 246;
© Nicolas Guiraud: p.11;
© Thierry Mesnard: pp.29 t., 31 b.;
© Nicolas Vereecken: pp.37, 54-55;
© Emmanuel de Bettignies: pp.76 b., 77 t.;
© Perrine Hervé-Gruyer: pp.81, 89, 90 b., 100 t., 116 t., 117, 121 b.l.;
© Rose Hervé-Gruyer: pp.99;
© D. R.: pp.112 b., 161 b., 207 t. and b.;
© Laure Chavanne: p.118 t. and b.;
© Pauline de Voghel: pp.121 t., 140, 145 m. and b., 192 b., 208 m., 230 b.;
© Max Riché: pp.125 b., 166 b., 201 b., 204 b., 240, 241 t., 224 t.-l., 247, 250;
© Xavier Dubois: p.134;
© Philippe Lemarchand: p.148;
© Gilles Degroote: pp.150, 151 t., 152;
© Andrea Sham: p.151 b.;
© Lila Hervé-Gruyer: p.153 t., 245 b.;
© Martin Egbert: pp.155 b.;
© Frédéric Sauvadet: p.173, 174 t., 175, 194 t.;
© Gauthier Chapelle: pp.201 t., 210, 211, 212 b.;
© Antoine Agoudjian: pp.230 t., 231 b.;
© Julien Vanlerberghe: p.205 b.;
© Olivier Marchant: p.217;
© Christophe Jammet: pp.241 b., 242 t., 245 m.;
© Babeth Aloy: p.243 t.

INDEX

PRAISE FOR THE BOOK

Perrine and Charles Hervé-Gruyer continue to deliver their powerful blend of deeply considered philosophy and practical guidance. If you are new to this world, you could not wish for a better lodestar. Even for those who think they know this subject well, I guarantee this book will bring a new perspective to your thinking. From the intricacies of seed saving to horse powered cultivation, through to business and market planning, if only I had had this beautifully illustrated book as I was starting out 30 years ago.

Ben Raskin,
Agroforester, Author and Soil Association Head of Agroforestry

This book masterfully combines practical and theoretical knowledge rooted in the cherished traditions of family farming on a micro scale.

From seed harvesting and propagation to effective pest management, finding the right farm to how to sell your produce, the authors cover every essential aspect of creating your own microfarm. They have crafted a textbook for the microfarm owner that's rich in wisdom, seamlessly incorporating traditional techniques like animal drafting and thoughtful tool maintenance. What truly sets this volume apart is the profound recognition of the human being as an integral part of the ecosystem, seamlessly woven throughout the text.

This book is not just informative; it is a delightful read that brims with fascinating and useful information. Whether you're a already a small scale farmer or a curious newcomer, this book is sure to inspire and empower you on your journey toward sustainable agriculture. An absolute treasure!

Rachel Phillips,
Director, Apricot Centre CIC

Tapping into the wisdom of nature is one of the best ways to maximize farm profits and efficiency. Charles and Perrine Hervé-Gruyer show pathways to making use of ecological principles for practical farm management in a digestible way for growers everywhere.

Zach Loeks,
Edible Ecosystem Designer and Author of *The Garden Tool Handbook*, *The Two-wheel Tractor Handbook* and *The Permaculture Market Garden*

There are two edges to making a regenerative small farm really work for you. There is knowing what you want and the practical, step by step, enacting of this, but there is also a belief in magic, beauty and the power of working with nature. This book spins together both songs in a way that is a gift to the reader. The work of Charles and Perrine at Bec Hellouin Farm has been instrumental in our developing of Henbant. We visited early in our own journey and absorbed what we could; this book gives so much more. How powerful to have all this inspiration and knowledge in this book. I recommend it to all who seek a meaningful and practical life in relation to the earth.

Matt Swarbrick,
Cofounder of Henbant Permaculture Farm

This is an important book. Highly recommended.

Kim Stoddart,
Amateur Gardening magazine Editor and Author of *The Climate Change Resilient Vegetable Garden*